Everyday Mathematics®

The University of Chicago School Mathematics Project

STUDENT MATH JOURNAL
VOLUME 1

Mc
Graw
Hill
Education

The University of Chicago School Mathematics Project

Max Bell, Director, *Everyday Mathematics* First Edition; James McBride, Director, *Everyday Mathematics* Second Edition; Andy Isaacs, Director, *Everyday Mathematics* Third, CCSS, and Fourth Editions; Amy Dillard, Associate Director, *Everyday Mathematics* Third Edition; Rachel Malpass McCall, Associate Director, *Everyday Mathematics* CCSS and Fourth Editions; Mary Ellen Dairyko, Associate Director, *Everyday Mathematics* Fourth Edition

Authors
Robert Balfanz*, Max Bell, John Bretzlauf, Sarah R. Burns**, William Carroll*, Amy Dillard, Robert Hartfield, Andy Isaacs, James McBride, Kathleen Pitvorec, Denise A. Porter‡, Peter Saecker, Noreen Winningham†

*First Edition only
**Fourth Edition only
†Third Edition only
‡Common Core State Standards Edition only

Fourth Edition Grade 5 Team Leader
Sarah R. Burns

Writers
Melanie S. Arazy, Rosalie A. DeFino, Allison M. Greer, Kathryn M. Rich, Linda M. Sims

Open Response Team
Catherine R. Kelso, Leader; Emily Korzynski

Differentiation Team
Ava Belisle-Chatterjee, Leader; Martin Gartzman, Barbara Molina, Anne Sommers

Digital Development Team
Carla Agard-Strickland, Leader; John Benson, Gregory Berns-Leone, Juan Camilo Acevedo

Virtual Learning Community
Meg Schleppenbach Bates, Cheryl G. Moran, Margaret Sharkey

Technical Art
Diana Barrie, Senior Artist; Cherry Inthalangsy

UCSMP Editorial
Don Reneau, Senior Editor; Rachel Jacobs, Elizabeth Olin, Kristen Pasmore, Loren Santow

Field Test Coordination
Denise A. Porter, Angela Schieffer, Amanda Zimolzak

Field Test Teachers
Diane Bloom, Margaret Condit, Barbara Egofske, Howard Gartzman, Douglas D. Hassett, Aubrey Ignace, Amy Jarrett-Clancy, Heather L. Johnson, Jennifer Kahlenberg, Deborah Laskey, Jennie Magiera, Sara Matson, Stephanie Milzenmacher, Sunmin Park, Justin F. Rees, Toi Smith

Digital Field Test Teachers
Colleen Girard, Michelle Kutanovski, Gina Cipriani, Retonyar Ringold, Catherine Rollings, Julia Schacht, Christine Molina-Rebecca, Monica Diaz de Leon, Tiffany Barnes, Andrea Bonanno-Lersch, Debra Fields, Kellie Johnson, Elyse D'Andrea, Katie Fielden, Jamie Henry, Jill Parisi, Lauren Wolkhamer, Kenecia Moore, Julie Spaite, Sue White, Damaris Miles, Kelly Fitzgerald

Contributors
John Benson, Jeanne Di Domenico, James Flanders, Fran Goldenberg, Lila K. S. Goldstein, Deborah Arron Leslie, Sheila Sconiers, Sandra Vitantonio, Penny Williams

Center for Elementary Mathematics and Science Education Administration
Martin Gartzman, Executive Director; Meri B. Fohran, Jose J. Fragoso, Jr., Regina Littleton, Laurie K. Thrasher

External Reviewers

The *Everyday Mathematics* authors gratefully acknowledge the work of the many scholars and teachers who reviewed plans for this edition. All decisions regarding the content and pedagogy of *Everyday Mathematics* were made by the authors and do not necessarily reflect the views of those listed below.

Elizabeth Babcock, California Academy of Sciences; Arthur J. Baroody, University of Illinois at Urbana-Champaign and University of Denver; Dawn Berk, University of Delaware; Diane J. Briars, Pittsburgh, Pennsylvania; Kathryn B. Chval, University of Missouri–Columbia; Kathleen Cramer, University of Minnesota; Ethan Danahy, Tufts University; Tom de Boor, Grunwald Associates; Louis V. DiBello, University of Illinois at Chicago; Corey Drake, Michigan State University; David Foster, Silicon Valley Mathematics Initiative; Funda Gönülateş, Michigan State University; M. Kathleen Heid, Pennsylvania State University; Natalie Jakucyn, Glenbrook South High School, Glenview, IL; Richard G. Kron, University of Chicago; Richard Lehrer, Vanderbilt University; Susan C. Levine, University of Chicago; Lorraine M. Males, University of Nebraska-Lincoln; Dr. George Mehler, Temple University and Central Bucks School District, Pennsylvania; Kenny Huy Nguyen, North Carolina State University; Mark Oreglia, University of Chicago; Sandra Overcash, Virginia Beach City Public Schools, Virginia; Raedy M. Ping, University of Chicago; Kevin L. Polk, Aveniros LLC; Sarah R. Powell, University of Texas at Austin; Janine T. Remillard, University of Pennsylvania; John P. Smith III, Michigan State University; Mary Kay Stein, University of Pittsburgh; Dale Truding, Arlington Heights District 25, Arlington Heights, Illinois; Judith S. Zawojewski, Illinois Institute of Technology

Note
Many people have contributed to the creation of *Everyday Mathematics*. Visit http://everydaymath.uchicago.edu/authors/ for biographical sketches of *Everyday Mathematics* 4 staff and copyright pages from earlier editions.

www.everydaymath.com

Copyright © McGraw-Hill Education

All rights reserved. No part of this publication may be reproduced or distributed in any form or by any means, or stored in a database or retrieval system, without the prior written consent of McGraw-Hill Education, including, but not limited to, network storage or transmission, or broadcast for distance learning.

Send all inquiries to:
McGraw-Hill Education
8787 Orion Place
Columbus, OH 43240

ISBN: 978-0-02-143099-4
MHID: 0-02-143099-3

Printed in the United States of America.

15 16 17 18 LMN 25 24 23 22

Contents

Unit 2

Unit 3

Unit 4

Activity Sheets

Welcome to *Fifth Grade Everyday Mathematics*

This year in math class you will continue to build on the mathematical skills and ideas you have learned in previous years. You will learn new mathematics and think about the importance of mathematics in your life now and how math will be useful to you in the future. Many of the new ideas you learn this year will be ones that your parents, or even your older brothers and sisters, may not have learned until much later than fifth grade. The authors of *Everyday Mathematics* believe that today's fifth graders are able to learn more and do more than fifth graders in the past. They think that mathematics is fun and they think you will find it enjoyable too.

Here are some of the things you will do in *Fifth Grade Everyday Mathematics:*

- Extend your understanding of place value to decimals and use what you learn to explain how our place-value system works.

- Review and extend your skills doing arithmetic, using a calculator, and thinking about problems and their solutions. You will add, subtract, multiply, and divide whole numbers and decimals.

- Use your knowledge of fractions and operations to compute with fractions. You will think about how adding, subtracting, multiplying, and dividing fractions is similar to and different from doing the same computations with whole numbers and decimals.

- Explore the concept of volume. You will learn how volume differs from other measurements you have studied. You will find the volume of 3-dimensional figures in multiple ways, and you will develop strategies for finding the volume of rectangular prisms. Look at journal page 2. Without telling anyone, write the number one hundred twelve in the top right-hand corner of the page.

- Learn about coordinate grids and find out how graphing can help you solve mathematical and real-world problems.

- Deepen your understanding of 2-dimensional figures, their attributes, and how different 2-dimensional figures are related to each other.

We want you to become better at using mathematics so you can better understand your world. We hope you enjoy the activities in *Fifth Grade Everyday Mathematics* and that they help you appreciate the beauty and usefulness of mathematics in your daily life.

Student Reference Book
Scavenger Hunt

Solve the problems on this page and page 3. Use your *Student Reference Book* to find information about each problem. Record the page numbers.

	Problem Points	Page Points

1 5 meters = _____ centimeters

 page _____

2 300 mm = _____ cm

SRB page _____

3 Solve.

(15 − 4) * 3 = _____

25 + (47 − 18) = _____

SRB page _____

4 Write the value of the 5 in each of the following numbers.

9,652 _____

15,690 _____

1,052,903 _____

SRB page _____

5 Name two fractions equivalent to $\frac{4}{6}$.

_____ and _____

SRB page _____

6 460 ÷ 5 = _____

 page _____

	Problem Points	Page Points

7   **a.** What is the definition of a trapezoid?   _____ _____

_____

_____

    b. Draw two different trapezoids.

    SRB   page _____

8   What materials do you need to play *Name That Number*?   _____ _____

_____

_____

SRB   page _____

Record your scavenger hunt scores in the table below. Then calculate the totals.

Problem Number	Problem Points	Page Points	Total Points
1			
2			
3			
4			
5			
6			
7			
8			
Total Points			

Math Boxes

① Next to each *Student Reference Book* icon in Problems 2–5, write the SRB page numbers where you can find information about each problem.

② Solve.

   a. (25 − 5) * 4 = _____

   b. 25 − (5 * 4) = _____

   _____

③ Complete.

   a. 1 foot = _____ inches

   b. A person who is 6 feet tall is

   _____ inches tall.

   c. 1 yard = _____ feet

   d. A person who ran 300 yards

   ran _____ feet.

   _____

④ Without calculating, circle ALL of the expressions that are greater than 2 + 8.

   A. 4 + (2 + 8)

   B. 2 + 8 − 5

   C. 2 + 8 + 10

   D. 8 + 2

   _____

⑤ **Writing/Reasoning** Explain how you solved Problem 2b.

   _____

   _____

   _____

   _____

   _____

   _____

Areas of Rectangles

Write two important facts that you learned about *area* after reading
Student Reference Book, page 221.

SRB
221

① _____

_____

② _____

_____

In Problems 3 and 4 each grid square is 1 square unit.
Find the area of each rectangle. Don't forget to include a unit.

③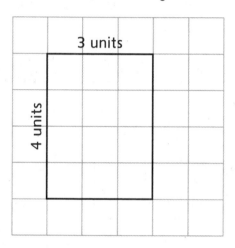

3 units

4 units

④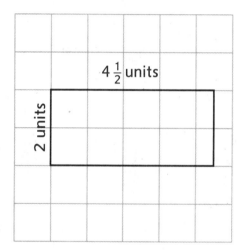

$4\frac{1}{2}$ units

2 units

Area: _____   Area: _____

⑤ Think about how you found the area of the rectangle in Problem 3 and how you found the
area of the rectangle in Problem 4. What was the same? What was different? Record your
thoughts. Be prepared to share them with the class.

_____

_____

_____

_____

5

Finding Areas of Rectangles

Find the area of each rectangle.
Write a number sentence to show your thinking.

SRB
221,
224-225

1

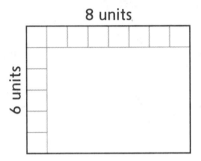

8 units

6 units

Area = _____ square units

_____

(number sentence)

2

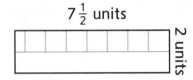

7$\frac{1}{2}$ units

2 units

Area = _____ square units

_____

(number sentence)

3

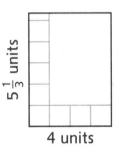

5$\frac{1}{3}$ units

4 units

Area = _____ square units

_____

(number sentence)

Try This

4

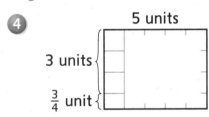

5 units

3 units

$\frac{3}{4}$ unit

Area = _____ square units

_____

(number sentence)

5 Explain the strategy you used to find the area of the rectangle in Problem 3. Use words like *row, column, square unit,* and *partial square* to help make your thinking clear.

_____

_____

_____

6

Math Boxes

1 Place parentheses in a different place in each problem. Then solve.

6 * 3 + 8 = _____

6 * 3 + 8 = _____

SRB
42

2 Find the area of the rectangle.

2 units

4 units

Area = _____ square units

SRB
224-225

3 Write an expression for the number story.

Alexa had 3 boxes of 10 shells. She bought shells at a store to double her collection.

_____

SRB
38, 44

4 Complete the table.

Feet	Inches
1	
10	
	60
	42

SRB
215-216,
328

5 Jamir has 137 dollars in the bank and 25 dollars at home. Ricky has twice as much money as Jamir.

Circle the expression that represents Ricky's money.

a. 2 * 137 + 25

b. 2 * (137 + 25)

SRB
42, 46

6 Give the value of the 4 in each number.

a. **4**2,671 _____

b. 6**4**,671 _____

c. 62,**4**71 _____

d. 62,6**4**1 _____

e. 62,67**4** _____

SRB
66-67

7

Area in Two Units

Noah has a piece of paper with an area of 1 square foot. He plans to paint a design on it using smaller squares, each with an area of 1 square inch. Below is the plan for his design.

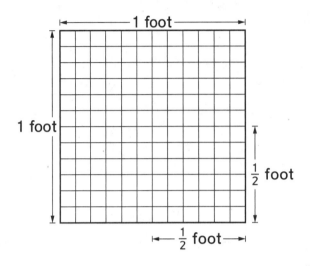

1. Use the picture to find the number of smaller squares (with an area of 1 square inch) that Noah will use in his design.

    He will use _____ square inches in his design.

2. Noah wants to make a second, smaller design in a square with a side length of $\frac{1}{2}$ foot. Use the picture to find the number of squares (with an area of 1 square inch) that Noah will use in the second design.

    He will use _____ square inches in his second design.

Math Boxes

1 Where in the *Student Reference Book* would you look to find the definition of *area*?

Circle the best answer.

a. Table of contents

b. Index

c. Glossary

d. Games section

e. All of the above

2 Solve.

a. $(4 * 12) + 8 =$ _____

b. _____ $= 32 / (16 \div 2)$

c. _____ $= (32 \div 8) * 2$

SRB
42

3 Draw lines to match each measurement with its equivalent.

a. 1 cm   1,000 m

b. 1 km   100 cm

c. 1 m $\frac{1}{1,000}$ m

d. 1 mm $\frac{1}{100}$ m

SRB
213, 328

4 Two friends were playing a game and recorded their scores below.

Player 1: 42 + 51 points

Player 2: 4 + (42 + 51) points

Who has more points? _____

SRB
46

5 **Writing and Reasoning** Did you have to calculate the scores to find out who had more points in Problem 4? Why or why not?

_____

_____

_____

SRB
46

9

A Tiling Strategy

Math Message

SRB
226

1 In Lesson 1-3 you found that 4 squares with a side length of $\frac{1}{2}$ foot fit into 1 square foot. How many squares with a side length of $\frac{1}{3}$ foot do you think would fit into 1 square foot? Use the pictures below to help. Be ready to explain how you found your answer.

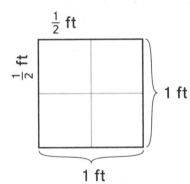

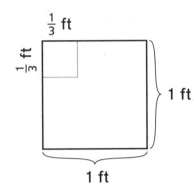

_____ squares with a side length of $\frac{1}{3}$ foot fit into one square foot.

---

2 Roger's shower is $2\frac{2}{3}$ feet wide and 3 feet long. He is going to cover the floor of the shower with square tiles that measure $\frac{1}{3}$ foot on each side.

a. How many tiles will Roger need to cover the shower floor? Use the picture to help.

_____ tiles

b. How many of Roger's tiles does it take to cover 1 square foot?

_____ tiles

c. Use your answers to Parts a and b to find the area of Roger's shower floor in square feet.

_____ square feet

_____
(number sentence)

$2\frac{2}{3}$ ft

3 ft

$\frac{1}{3}$ ft

$\frac{1}{3}$ ft

3 Summarize the strategy you used to find the area of Roger's shower floor.

_____

_____

Solving Area Problems

1 Anna is covering the top of her jewelry box with glass tiles that are $\frac{1}{2}$ inch long and $\frac{1}{2}$ inch wide. The top of the jewelry box is $3\frac{1}{2}$ inches by 2 inches.

SRB
226

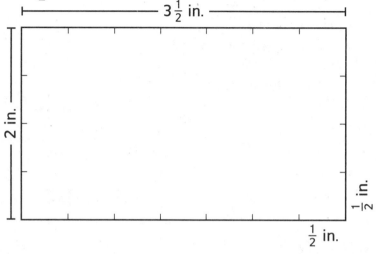

a. How many tiles will she need to cover the top of the box? Use the picture to help.

_____ tiles

b. How many of Anna's tiles does it take to cover 1 square inch?

_____ tiles

c. Use your answers to Parts a and b to find the area of the top of the jewelry box in square inches. _____ square inches

_____
(number sentence)

2 Deshawn is covering a 4-yard by $1\frac{3}{4}$-yard section of his bedroom wall with decorative tiles. The tiles are $\frac{1}{4}$ yard by $\frac{1}{4}$ yard.

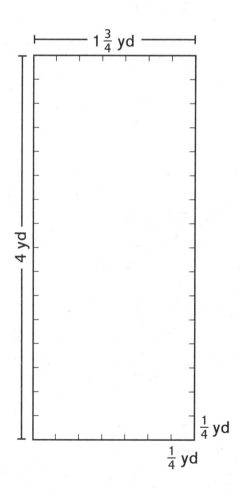

a. How many tiles will Deshawn need to cover the section of the wall? Use the picture to help.

_____ tiles

b. How many tiles would it take to cover 1 square yard?

_____ tiles

c. Use your answers to Parts a and b to find the area in square yards of the section of the wall that Deshawn is decorating.

_____ square yards

_____
(number sentence)

11

Math Boxes

Math Boxes

① Place parentheses in a different place in each problem. Then solve.

4 * 8 − 2 = _____

4 * 8 − 2 = _____

SRB
42

② Find the area of the rectangle.

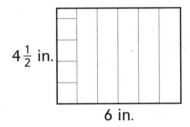

$4\frac{1}{2}$ in.

6 in.

Area = _____ square inches

SRB
224-225

③ Write an expression for the number story.

Juno earned 140 points in a game. He lost half of the points and then earned 160 more.

_____

SRB
38, 44

④ Complete.

a. 60 inches = _____ feet

b. 3 yards = _____ feet

c. 4 meters = _____ cm

d. 2 miles = _____ yards

SRB
215-216,
328

⑤ Expression A: Expression B:
    458 − 12   25 + (458 − 12)

Check all that apply.

☐ B is greater than A.

☐ B is twice as much as A.

☐ B is 25 more than A.

SRB
42, 46

⑥ Write a 5-digit number that has

7 in the ones place,

8 in the hundreds place,

4 in the ten-thousands place,

and 0 in all other places.

____ ____, ____ ____ ____

SRB
66-67

Comparing Volume

1 Create Cylinders A and B from half-sheets of paper. What could you measure about these cylinders?

SRB
230

_____

_____

2 Which cylinder do you think has a greater volume? Explain your answer.

_____

_____

_____

3 Test your prediction. Which cylinder has a greater volume? Explain your answer.

_____

_____

_____

_____

4 Think about all the attributes you listed in Problem 1.

   a. How are the cylinders different?

   _____

   _____

   b. How are the cylinders the same?

   _____

   _____

13

Math Boxes

Math Boxes

1 Solve.

    a. $2 * (14 + 6) =$ _____

    b. $(21 / 3) + 14 =$ _____

    c. $(10 * 8) - 20 =$ _____

SRB
42

2 Name three objects that have the attribute of volume.

_____

_____

_____

SRB
230

3 Jo's closet is 6 ft wide and $1\frac{1}{2}$ ft deep. Find the area of the closet floor.

$1\frac{1}{2}$ ft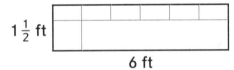

6 ft

Area = _____ square feet

SRB
224-225

4 Which numerical expression shows the following calculation? Fill in the circle next to the best answer.

Add seven and three, then multiply by 6.

Ⓐ $(6 * 7) + 3$

Ⓑ $(7 + 3) * 6$

Ⓒ $7 + 3 + 6$

SRB
42, 45

5 **Writing/Reasoning** Explain how you found the area of Jo's closet floor in Problem 3.

_____

_____

_____

_____

SRB
224-225

14

Packing Prisms to Measure Volume

1. Use pattern blocks to measure the volume of your rectangular prism. Record your results below.

   SRB
   230

   Our prism has a volume of about _____ **square** pattern blocks.

   Our prism has a volume of about _____ **triangle** pattern blocks.

   Our prism has a volume of about _____ **hexagon** pattern blocks.

2. What was important to remember as you packed the prism with pattern blocks so you could measure volume as accurately as possible?

   _____

   _____

3. Which pattern block did you need the **most** of to fill your prism? Why?

   _____

   _____

4. Which pattern block did you need the **least** of to fill your prism? Why?

   _____

   _____

5. What other objects could you use to fill the prism?

   _____

   _____

6. What 3-dimensional shape do you think would be easiest to pack tightly into a rectangular prism without gaps or overlaps? Why do you think so?

   _____

   _____

   _____

More Areas of Rectangles

DATE   TIME

Find the area of each rectangle.

1

5 in.

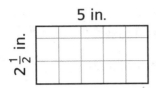

$2\frac{1}{2}$ in.

Area = _____ square inches

2

$4\frac{1}{3}$ m

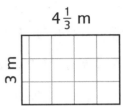

3 m

Area = _____ square meters

3

6 yd

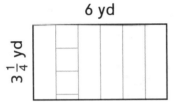

$3\frac{1}{4}$ yd

Area = _____ square yards

4

$2\frac{1}{5}$ cm

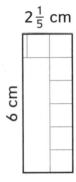

6 cm

Area = _____ square centimeters

Try This

5 **a.** What is the area of this rectangle if
the sides of the squares are each 1 unit long?

Area = _____ square units

b. What is the area of this rectangle if the
sides of the squares are each $\frac{1}{3}$ unit long?

Area = _____ square units

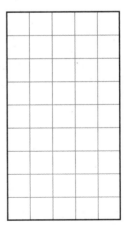

Math Boxes

1 Box A has a volume of 152 beans. If Box B has a greater volume than Box A, which could be the volume of Box B?

Choose the best answer.

◯ 25 beans

◯ 100 beans

◯ 125 beans

◯ 200 beans

SRB
230

2 What is the area of a rectangle that is $2\frac{1}{4}$ in. wide and 4 in. long?

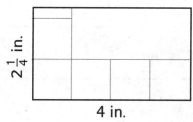

$2\frac{1}{4}$ in.

4 in.

Area = _____ square inches

SRB
224-225

3 Which expressions are less than $16 - 8$?

Circle ALL that apply.

A. $(16 - 8) * 2$

B. $(16 - 8) - 2$

C. $(16 - 8) \div 2$

D. $(16 - 8) + 2$

SRB
46

4 Complete.

a. 1 hour = _____ minutes

b. $3\frac{1}{2}$ hours = _____ minutes

c. $\frac{3}{4}$ hour = _____ minutes

d. _____ hours = 240 minutes

SRB
215-216,
328

5 Write an expression for each statement.

Add 12 and 8 and multiply the sum by 3.

_____

Subtract 10 from the product of 6 and 8.

_____

SRB
42, 45

6 Josh had 10 goldfish and 2 guppies. Half of the fish are male. Write an expression that shows how many fish are female.

_____

SRB
42, 44

Math Boxes

17

Measuring Volume with Cubes

Record your estimates from the Math Message in the second column of the table. Record the actual number of cubes you used to fill the prism in the third column of the table.

SRB
231

Rectangular prism	Estimated Number of Cubes to Fill the Prism	Actual Number of Cubes to Fill the Prism
A		
B		
C		

The cubes in each rectangular prism below are the same size.
Each prism has at least one stack of cubes that goes up to the top.
Find the total number of cubes needed to completely fill each prism.

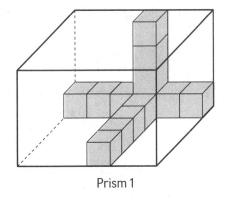

Prism 1

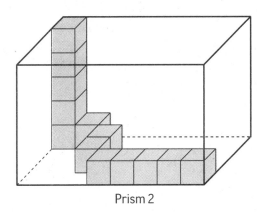

Prism 2

Cubes needed to fill Prism 1: _____ cubes

Volume of Prism 1: _____ cubic units

Cubes needed to fill Prism 2: _____ cubes

Volume of Prism 2: _____ units³

The cubes in each rectangular prism are the same size.
Each prism has at least one stack of cubes that goes up to the top.
Find the number of cubes needed to completely fill each prism.

SRB
231

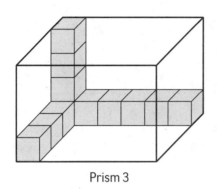

Prism 3

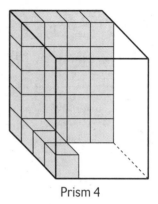

Prism 4

Cubes needed to fill Prism 3: _____ cubes

Volume of Prism 3: _____ cubic units

Cubes needed to fill Prism 4: _____ cubes

Volume of Prism 4: _____ units^3

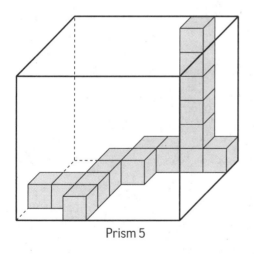

Prism 5

Try This

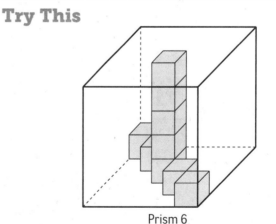

Prism 6

Cubes needed to fill Prism 5: _____ cubes

Volume of Prism 5: _____ units^3

Cubes needed to fill Prism 6: _____ cubes

Volume of Prism 6: _____ cubic units

Math Boxes

Math Boxes

① Solve.

   a. (4 * 3) + (9 / 3) = _____

   b. 4 * (2 + 7) − 6 = _____

   c. _____ = (10 + 8) / (54 / 9)

SRB
42

② If you want to know how many boxes of markers will fit in a drawer, do you need to know the length, area, or volume of the drawer?

_____

SRB
218-219,
221, 230

③ Find the area of a table that is 3 feet wide and $2\frac{1}{2}$ feet long.

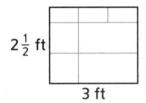

$2\frac{1}{2}$ ft

3 ft

Area = _____ square feet

SRB
224-225

④ Write the following expressions using numbers.

   a. seven times the sum of 6 and 4

      _____

   b. the difference of sixteen and eight, divided by two

      _____

SRB
42, 45

⑤ **Writing/Reasoning** How did you decide which type of measurement you would need to know in Problem 2?

_____

_____

_____

SRB
218-219,
221, 230

Converting Measurements

Solve. If necessary, look up measurement equivalents in the *Student Reference Book*.

SRB
215-216,
328

1 Record measurement equivalents in the 2-column tables below.

Minutes	Seconds
5	
10	
15	
20	
25	

Kiloliters	Liters
	5,000
	1,500
$2\frac{1}{2}$	
4	
	3,500

Miles	Yards
1	1,760
2	
3	
4	
5	

2 Students recorded their running distances over the weekend using different units. Complete the table to convert them to the same units.

	Kilometers	Meters
Jason	3	
Kayla		4,500
Lohan	$2\frac{1}{2}$	
Malik		5,000
Jada	$3\frac{1}{2}$	

3 Jordan needed to convert measurements for his recipes. Fill in the blanks.

2 quarts milk = _____ cups

_____ cups pasta = 2 pints

32 oz flour = _____ lb

$2\frac{1}{2}$ cups water = _____ fl oz

8 cups rice = _____ pints

$\frac{1}{2}$ cup oil = _____ tbs

4 Mahalia is making cloth napkins. She bought fabric that is 12 inches wide and 6 yards long.

How many napkins can Mahalia make if each napkin is 1 foot by 1 foot? _____ napkins

Cube-Stacking Problems Using Layers

Complete the table for each rectangular prism.

SRB
232

Rectangular Prism	Number of Cubes in 1 Layer	Number of Layers	Total Number of Cubes That Fill the Prism	Volume of the Prism
G				_____ cubic units
H				_____ cubic units
I				_____ cubic units
J				_____ cubic units
K				_____ cubic units
L				_____ cubic units

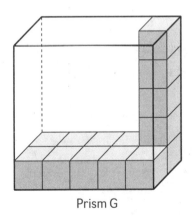

Prism G

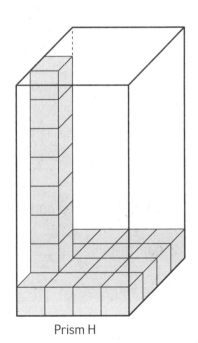

Prism H

22

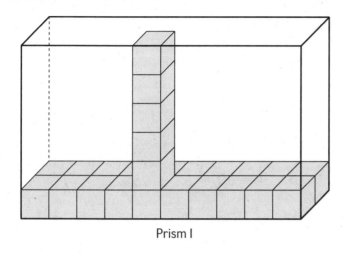

Prism I

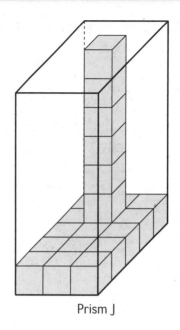

Prism J

SRB
232

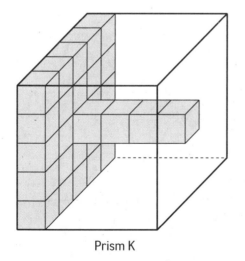

Prism K

Try This

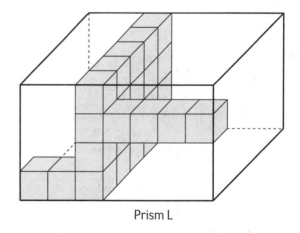

Prism L

Math Boxes

① What is the value of the 8 in the following numbers?

a. 1,384 _____

b. 8,294 _____

c. 418 _____

d. 6,897 _____

SRB
66-67

② Solve.

a. $3 * 10 =$ _____

b. $3 * 100 =$ _____

c. $3 * 1,000 =$ _____

d. $30 * 10 =$ _____

SRB
95

③ Write four multiples of 4.

_____ , _____ , _____ , _____

SRB
72

④ Solve.

a. $\begin{array}{r} 2\ 3 \\ *\quad\ 3 \\ \hline \end{array}$ b. $\begin{array}{r} 2\ 4\ 2 \\ *\quad\quad 2 \\ \hline \end{array}$

SRB
100-101,
104

⑤ Solve.

How many 10s in 30? _____

How many 10s in 300? _____

How many 7s in 21? _____

How many 7s in 210? _____

SRB
106

⑥ Solve.

$50 + 6 =$ _____

$300 + 20 =$ _____

$200 + 50 + 6 =$ _____

SRB
70

Finding Volume Using Formulas

Use a formula to find the volume of each prism. Record the formula you used.

①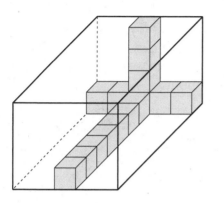

Volume: _____

Formula: _____

②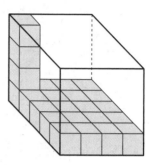

Volume: _____

Formula: _____

③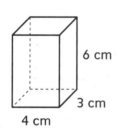

6 cm

3 cm

4 cm

Volume: _____

Formula: _____

④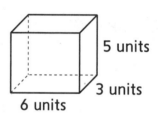

5 units

3 units

6 units

Volume: _____

Formula: _____

⑤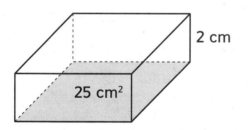

2 cm

25 cm²

Volume: _____

Formula: _____

Try This

⑥ A rectangular prism has a volume of 36 cubic units. Write two different possible sets of dimensions for the prism.

Set 1:

length = _____

width = _____

height = _____

Set 2:

length = _____

width = _____

height = _____

25

Writing and Interpreting Expressions

Write an expression that models the calculation described in words.

1 The sum of 13 and 12, which is then multiplied by 2

_____

2 Divide 16 by 4 and add the sum of 3 and 8 to the quotient.

_____

3 Multiply 12 and 6 and divide the product by 9.

_____

Without calculating, circle the expression with the greater value.

4 3 * (126 + 12)   6 * (126 + 12)

5 (18 − 8) / 2   (18 − 8) / 5

6 Explain how you knew which expression had a greater value in Problem 5.

_____

_____

_____

7 Ivan was playing a video game. He had 1,300 points and on the next level earned 120 more. Then he lost 12 points. When his turn ended, his score doubled. Write an expression that shows the number of points Ivan has at the end of his turn.

_____

8 Write a situation that can be modeled by the expression 6 * (24 − 5).

_____

_____

_____

26

Math Boxes

Math Boxes

① What is the volume of the prism?

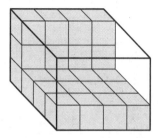

Volume = _____ cubic units

SRB
231-232

② Raoul says he can find the area of a rectangle that is $1\frac{1}{2}$ in. by 3 in. using addition. Write a number model that Raoul could use to find the area of the rectangle.

Area = _____

SRB
224-225

③ This prism is made of unit cubes. Use $V = B * h$ to find the volume.

Volume = _____ * _____ = _____ cubic units
       area of   height
       base

SRB
231-233

④ Insert parentheses to make the equations true.

a. $2 + 3 * 4 = 20$

b. $4 * 5 - 3 = 8$

SRB
42

⑤ **Writing/Reasoning** How did you figure out the volume of the prism in Problem 1?

_____

_____

_____

_____

SRB
231-232

27

Converting Cubic Units

1 Is a cubic inch larger or smaller than a cubic centimeter? How do you know?

SRB
235

_____

_____

2 List objects with volumes you might measure in cubic inches.

_____

3   **a.** How many cubic inches do you think are in a cubic foot? _____

b. How many inches are in a foot? _____

c. How many square inches are in a square foot?

_____ square inches

How did you find your answer?

_____

_____

d. How many cubic inches are in a cubic foot?

_____ cubic inches

How did you find your answer?

_____

_____

_____

4 List objects with volumes you might measure in cubic feet.

_____

_____

Converting Cubic Units (continued)

5 How many cubic feet are in a cubic yard?

_____ cubic feet

How did you find your answer?

_____

_____

_____

6 List objects with volumes you might measure in cubic yards.

_____

_____

7 Deena's family has a freezer that is 2 yards in width, 1 yard in length, and 1 yard in height.

a. What is the volume of the freezer?

_____ cubic yards

b. How many cubic feet of food will fit in the freezer?

_____ cubic feet

How did you find your answer?

_____

_____

c. Do you think cubic yards or cubic feet are better units to measure the volume of the freezer? Why?

_____

_____

More Cube-Stacking Problems

The cubes in each rectangular prism are the same size. Find the volume of each prism.

SRB
231-233

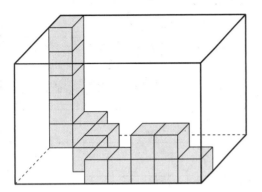

Volume = _____ cubic units

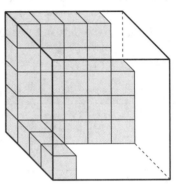

Volume = _____ cubic units

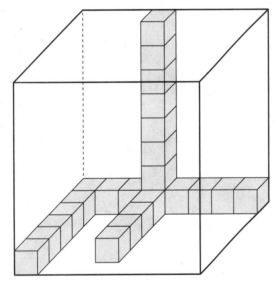

Volume = _____ cubic units

Try This

4 The stack of cubes represents the height of a rectangular prism.

Draw the outline of a rectangular base on the grid paper so that the volume of the prism would be 60 cubic units.

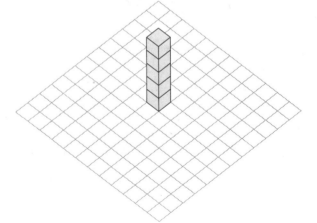

What is the area of the base you drew?

_____ square units

30

Math Boxes

1 Can A has a volume of 25 beans. Can B holds twice as many beans as Can A. What is the volume of Can B?

Volume = _____ beans

SRB
230

2 Find the area of a rectangle that is $1\frac{1}{2}$ feet by 2 feet.

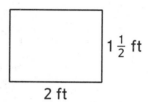

$1\frac{1}{2}$ ft

2 ft

Area = _____ square feet

SRB
224-225

3 If 329 + 671 = 1,000, what is 2 * (329 + 671)?

_____

SRB
46

4 Complete.

a. 100 yd = _____ ft

b. 2 mi = _____ ft

c. _____ in. = 2 yd

d. _____ ft = 30 in.

SRB
215-216, 328

5 Solve.

a. (14 + 2) / 8 = _____

b. (36 / 6) + (42 / 7) = _____

c. 3 * (50 + 20 + 30) = _____

SRB
42

6 Maria bought 5 tickets for $20 each. Each ticket had a $2 fee. Write an expression that shows how much Maria paid for the tickets.

_____

SRB
42, 44

31

Estimating Volumes of Instrument Cases

In Problems 1–3, use the mathematical models to estimate the volumes of the instrument cases.

1 Trombone case

The volume of the trombone case is

about _____ in.³.

2 French horn case

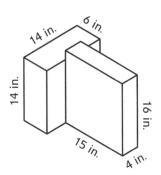

The volume of the French horn case is

about _____ in.³.

3 Xylophone case

 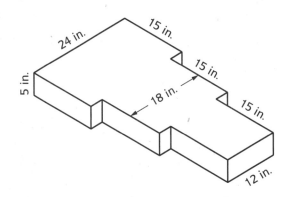

The volume of the xylophone case is about _____ in.³.

Try This

4 Asher needs to take the xylophone, the trombone, and the French horn with him to a band concert. His trunk has 13 cubic feet of cargo space. Can he fit all three cases in his trunk? Explain how you know.

_____

_____

_____

McGraw-Hill Education

Math Boxes

---

1 What is the volume of the prism?

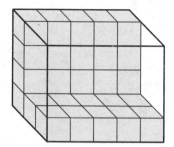

Volume = _____ cubic units

SRB
231-232

---

2 A doormat is 3 feet by $2\frac{1}{2}$ feet. What is the area of the doormat?

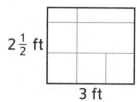

$2\frac{1}{2}$ ft

3 ft

Area = _____ ft²

SRB
224-225

---

3 This prism is made of unit cubes. Use $V = B \times h$ to find the volume.

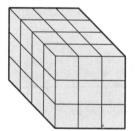

Volume = _____ × _____ = _____ cubic units
       area of   height
       the base

SRB
231-233

---

4 Insert parentheses to make the equations true.

a. $5 * 4 - 2 = 10$

b. $8 - 7 * 25 = 25$

c. $36 \div 6 - 5 = 36$

SRB
42

---

5 **Writing/Reasoning** What does the area of the base tell you about the number of cubes that fit in the prism in Problem 3?

_____

_____

_____

SRB
231-233

Understanding Grouping Symbols

Evaluate the following expressions.

SRB
38,
42-44

1 $10 * [13 + (12 - 7)] = $ _____

2   _____ $= \{(5 * 6) + 2\} / 4$

3   _____ $= \{13 + (2 * 1)\} * 3$

4 $64 / [20 - (4 * 3)] = $ _____

Insert grouping symbols to make the following number sentences true.

5 $4 = 4 * 6 - 2 + 3$

6 $300 \div 6 + 4 * 2 + 8 = 3$

7 $70 / 13 - 2 + 1 = 7$

8 $160 = 8 * 16 + 12 - 4 \div 2$

Write an expression that models the story. Then evaluate the expression.

9   Tommy had a bag of 100 balloons. He took out 2 red balloons and 1 blue balloon for each party favor. He created 12 party favors. How many balloons did Tommy have left?

Expression: _____

Answer: _____ balloons

10   A grocery store received a shipment of 100 cases of apple juice. Each case contained four 6-packs of cans. After inspection, the store found that 9 cans were damaged. How many cans were undamaged?

Expression: _____

Answer: _____ cans

Math Boxes

1 Jayden filled a tub with boxes of markers. He put 2 boxes in each layer. It took 4 layers to fill the tub. What is the volume of the tub?

Volume = _____ marker boxes

SRB
230, 232

2 Imani's room is 3 yards by $4\frac{1}{4}$ yards. She is getting new carpet.

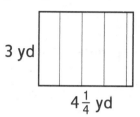

3 yd

$4\frac{1}{4}$ yd

Is 13 square yards of carpet enough to cover her room? _____

SRB
224-225

3 Write an expression with a value that is twice as much as $236 + 912$.

_____

SRB
45-46

4 Complete.

a. 1.5 km = _____ m

b. 36 in. = _____ yd

c. 3 m = _____ mm

d. 5 dm = _____ cm

SRB
215-216,
328

5 Solve.

a. $\left(\frac{1}{2} + \frac{1}{2}\right) * 5 =$ _____

b. $10 * (300 \div 30) =$ _____

c. $6 * \left(1\frac{1}{2} + 1\frac{1}{2}\right) =$ _____

SRB
42,
186-187

6 Milo has 3 dogs and 2 cats. Each pet eats 2 cups of food a day.

Which expression(s) below show how much food Milo's pets eat in one day?

Fill in the circle next to <u>all</u> that apply.

Ⓐ $(3 * 2) + 2$

Ⓑ $(3 + 2) * 2$

Ⓒ $3 + (2 + 2)$

Ⓓ $(3 * 2) + (2 * 2)$

SRB
42, 44

Math Boxes

Math Boxes: Preview for Unit 2

1 Use the numbers below to solve. Use each digit once.

7, 1, 0, 2, 9

a. Write the largest number you can.

_____

b. Write the smallest number you can. (Do not start with 0.)

_____

SRB
66-67

2 Solve.

a. $6 * 100 =$ _____

b. $6 * 10 =$ _____

c. $6 * 1,000 =$ _____

d. $60 * 10 =$ _____

SRB
95

3 Which of the following are all multiples of 6? Choose the best answer.

◯ 24, 72, 19, 36

◯ 12, 27, 42, 18

◯ 60, 48, 12, 24

◯ 6, 56, 42, 30

SRB
72

4 Solve.

a.   5 6
         *   4
         _____

b.   4 2 3
         *   6
         _____

SRB
100-101, 104

5 Solve.

$27 ÷ 9 =$ _____

$270 ÷ 9 =$ _____

_____ $= 42 ÷ 7$

_____ $= 420 ÷ 7$

SRB
106

6 Write the following number in expanded form:

$23,465 =$ _____

_____

SRB
70

36

Place-Value Relationships

Math Message

SRB
66-67,
70

1. What is the value of the 2 in the following numbers?

   2 _____

   23 _____

   230 _____

   2,300 _____

   23,000 _____

2. What is the value of the 6 in the following numbers?

   65,000 _____

   6,500 _____

   650 _____

   65 _____

   6 _____

---

3. Write these numbers in expanded form.

   a. 2,387,926 = _____

   _____

   b. 92,409,224 = _____

   _____

4. Write these numbers in standard notation.

   a. 4 [100,000s] + 5 [10,000s] + 0 [1,000s] + 3 [100s] + 6 [10s] + 2 [1s] = _____

   b. (9 * 10,000) + (3 * 1,000) + (4 * 100) + (9 * 10) + (1 * 1) = _____

   c. 3 ten-thousands + 2 thousands + 5 hundreds + 7 tens + 9 ones = _____

5. How does expanded form help you see the patterns in our place-value system?

   _____

   _____

   _____

37

Place-Value Relationships
(continued)

Lesson 2-1

DATE   TIME

SRB
66-67

6 Write the value of the 4 in each number.

    a. 348,621 _____

    b. 24,321 _____

    c. 624,876,712 _____

    d. 13,462 _____

    e. 463,295 _____

    f. 942 _____

7 Write the value of the identified digit. Then fill in the blank with "10 times" or "$\frac{1}{10}$ of."

    a. What is the value of 7 in 732? _____ In 7,328? _____

    The value of the 7 in 732 is _____ the value of 7 in 7,328.

    b. What is the value of 4 in 32,940? _____ In 32,904? _____

    The value of 4 in 32,940 is _____ the value of 4 in 32,904.

    c. What is the value of 2 in 30,275? _____ In 18,921? _____

    The value of 2 in 30,275 is _____ the value of 2 in 18,921.

    d. What is the value of 1 in 90,106? _____ In 21,000? _____

    The value of the 1 in 90,106 is _____ the value of 1 in 21,000.

8   **a.** Write a number where 5 is worth 500. _____

    b. Write a number where 5 is worth 10 times as much as the number you wrote in Part a.

    _____

    c. How did the position of the 5 change in your number in Part b?

    _____

9   **a.** Write a number where 3 is worth 30,000 and 2 is worth 20.

    _____

    b. Write a number where 3 is worth $\frac{1}{10}$ as much and 2 is worth 10 times as much as

    the number you wrote in Part a. _____

    c. How did the position of the 3 change in your number in Part b?

    _____

Finding Volumes of Rectangular Prisms

The cubes in each rectangular prism are the same size. Each prism has at least one stack of cubes that goes up to the top. Find the number of cubes needed to completely fill each prism. Then use the formula to help you write a number sentence that represents the volume.

SRB
231-233

①

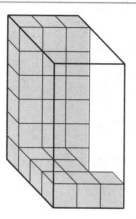

_____ cubes to fill the prism

$V = B * h$

_____ = _____ * _____

②

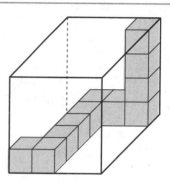

_____ cubes to fill the prism

$V = l * w * h$

_____ = _____ * _____ * _____

③

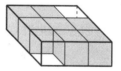

_____ cubes to fill the prism

$V = l * w * h$

_____ = _____ * _____ * _____

④

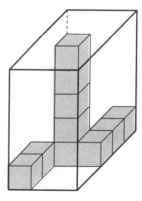

_____ cubes to fill the prism

$V = B * h$

_____ = _____ * _____

39

Math Boxes

1 Find the area of the rectangle.
Write a number sentence.

$1\frac{1}{4}$ cm
3 cm

Area = _____ cm²

_____
(number sentence)

SRB
224-225

2 What is the value of the 4 in the following numbers?

a. 42 _____

b. 420 _____

c. 4,200 _____

d. 42,000 _____

SRB
66-67

3 Complete.

a. 6 feet = _____ inches

b. _____ tons = 4,000 pounds

c. $\frac{1}{2}$ pound = _____ ounces

d. 9 yards = _____ feet

SRB
215-217,
328

4 Find the volume of the prism.

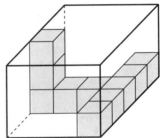

Volume = _____ cubic units

SRB
231-232

5 **Writing/Reasoning** How did you find the area of the rectangle in Problem 1?

_____

_____

_____

_____

SRB
224-225

Powers of 10 and Exponential Notation

SRB
68-69

Follow your teacher's instructions to complete this table.

Exponential Notation	Product of 10s	Standard Notation
10^6	10 * 10 * 10 * 10 * 10 * 10	1,000,000

Complete the number sentences.

1 $3 * 10^2 = 3 * 100 = $ _____

2 $7 * 10^9 = 7 * $ _____ $ = $ _____

3 $3 * 10^4 = $ _____ $ * $ _____ $ = $ _____

4 $25 * 10^3 = $ _____ $ * $ _____ $ = $ _____

5 _____ $ = 93 * $ _____ $ = 93,000,000$

6 The numbers in Problems 1–5 are the answers to the following questions. Fill in each blank with your best guess. Write your answer in exponential or standard notation.

   a. The distance from the sun to Earth is about _____ miles.

   b. The Statue of Liberty is about _____ feet tall.

   c. In 2014, the population of Earth was about _____ people.

   d. If you walked around Earth's equator, you would walk about _____ miles.

   e. Mount Everest, the highest mountain on Earth, is about _____ feet high.

Expanded Form with Powers of 10

Numbers in expanded form are written as addition expressions showing the value of each digit.

SRB
68-70

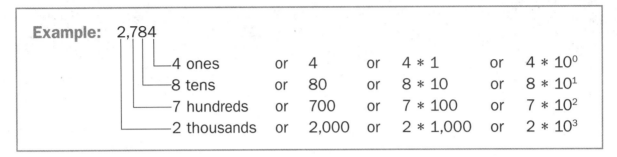

Example: 2,784

4 ones	or	4	or	4 * 1	or	4 * 10^0
8 tens	or	80	or	8 * 10	or	8 * 10^1
7 hundreds	or	700	or	7 * 100	or	7 * 10^2
2 thousands	or	2,000	or	2 * 1,000	or	2 * 10^3

2,784 can be written in expanded form in different ways.

- As an addition expression: 2,000 + 700 + 80 + 4

- As the sum of multiplication expressions involving powers of 10: (2 * 1,000) + (7 * 100) + (8 * 10) + (4 * 1)

- As the sum of multiplication expressions using exponents to show the powers of 10: (2 * 10^3) + (7 * 10^2) + (8 * 10^1) + (4 * 10^0)

1 **a.** Write 6,125 in expanded form as an addition expression.

_____

b. Write 6,125 in expanded form as the sum of multiplication expressions involving powers of 10.

_____

c. Write 6,125 in expanded form as the sum of multiplication expressions using exponents to show the powers of 10.

_____

2 Write each number in standard notation.

a. 12 * 10^5 _____   **b.** 4 * 10^8 _____

3 Write each number using exponential notation and powers of 10.

a. 30,000 _____   **b.** 4,200,000 _____

Solving a Real-World Volume Problem

Josef and his mother are renting a storage unit. They want to rent the largest unit. The dimensions of the available units are shown below. Calculate the volume of each unit. Write the formula you use and show your work. Circle the storage unit you think they should rent.

SRB
233-234

Storage Unit 1

8 ft   6 ft   5 ft

Formula: V = _____

The volume of Storage Unit 1 is _____ ft³.

Storage Unit 2

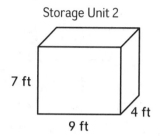

7 ft   4 ft   9 ft

Formula: V = _____

The volume of Storage Unit 2 is _____ ft³.

Storage Unit 3

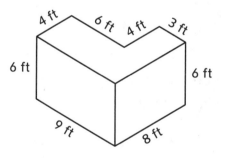

4 ft   6 ft   4 ft   3 ft
6 ft   6 ft
9 ft   8 ft

Formula: V = _____

The volume of Storage Unit 3 is _____ ft³.

Storage Unit 4

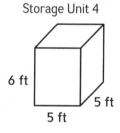

6 ft   5 ft   5 ft

Formula: V = _____

The volume of Storage Unit 4 is _____ ft³.

Math Boxes

Math Boxes

① Solve.

   a. $(24 \div 8) \times 4 =$ _____

   b. $4 + (15 / 3) =$ _____

   c. $[(6 + 4) \times 3] + 6 =$ _____

   d. $4 \times \{5 + (10 \div 2)\} =$ _____

SRB
42-43

② Write a 4-digit number with
4 in the hundreds place,
8 in the thousands place,
3 in the ones place,
and 7 in the tens place.

_____, _____ _____ _____

SRB
66-67

③ Write 23,436 in expanded form.

_____

_____

_____

_____

SRB
70

④ Find the volume of the prism. Use the
formula: $V = l \times w \times h$.

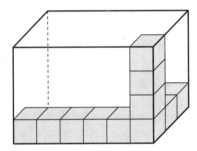

Volume = ____ × ____ × ____ = ____ units³

SRB
231-233

⑤ Write each power of 10 in exponential
notation.

   a. $10 \times 10 \times 10 =$ _____

   b. $10 \times 10 \times 10 \times 10 \times 10 =$ _____

   c. $10 \times 10 \times 10 \times 10 \times 10 \times 10 \times 10 =$

   _____

   d. $10 \times 10 \times 10 \times 10 \times 10 \times 10 \times 10 \times$

   $10 \times 10 =$ _____

SRB
68

⑥ Jonah's sister is 10 years old. Jonah is
8 years younger than twice his sister's
age. Write an expression for Jonah's age.

_____

SRB
42, 44

44

Estimating with Powers of 10

Use estimation to solve.

1. A hardware store sells ladders that extend up to 12 feet. The store's advertising says:

   Largest inventory in the country! If you put all our ladders end to end, you could climb to the top of the Empire State Building!

   The company has 295 ladders in stock. The Empire State Building is 1,453 feet tall.

   Is it true that the ladders would reach the top of the building? _____

   Explain how you solved the problem.

   _____

   _____

   _____

2. The school library received a donation of 42 boxes of books. Each box contains 15–18 books. The library has 10 empty bookshelves. Each bookshelf can hold up to 60 books.

   Does the library have enough shelf space for all the new books? _____

   Explain how you solved the problem.

   _____

   _____

   _____

3. Nishant is in charge of collecting cereal box tops to trade in for technology items for his school. He keeps the box tops in 38 folders. Each folder contains 80 box tops.

   Does Nishant have enough box tops to

   trade in for a printer? _____

   For a digital camera? _____

   For a tablet computer? _____

Item	Number of Box Tops Needed
Printer	1,500
Digital camera	3,500
Tablet computer	5,000

   Explain how you solved the problem.

   _____

   _____

   _____

45

Writing and Comparing Expressions

1   Write each statement as an expression using grouping symbols. Do not evaluate the expressions.

    a.   Find the sum of 13 and 7, then subtract 5. _____

    b.   Multiply 3 and 4 and divide the product by 6. _____

    c.   Add 138 and 127 and multiply the sum by 5. _____

    d.   Divide the sum of 45 and 35 by 10. _____

    e.   Add 300 to the difference of 926 and 452. _____

2   Compare the two expressions. Do not evaluate them. How are their values different?

    a. $2 * (489 + 126)$ and $489 + 126$

    _____

    _____

    b. $(367 \times 42) - 328$ and 367×42

    _____

    _____

3   Below are advertisements for two stores having sales on T-shirts.

    a.   Write an expression that represents the cost of two shirts at each store.

Shirts-R-Us	T-Shirt Mart
Half off T-shirts! Regular price: $14.00.	T-shirt Sale Price: $8.00.
Expression:	Expression:

    b.   Which store has the better deal? How do you know?

    _____

    _____

    _____

Math Boxes

Math Boxes

① Krista has a garden that is 5 yards by $2\frac{1}{2}$ yards and wants to cover it with compost. The compost is sold in bags that cover 5, 15, or 30 square yards. Which size bag should Krista buy?

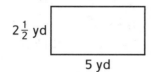

$2\frac{1}{2}$ yd

5 yd

Fill in the circle next to the best answer.

Ⓐ 5 square yards

Ⓑ 15 square yards

Ⓒ 30 square yards

SRB
224-225

② What is the value of the 8 in the following numbers?

a. 38 _____

b. 382 _____

c. 832 _____

d. 8,432 _____

e. 85,432 _____

SRB
66-67

③ Complete.

a. 5 m = _____ mm

b. _____ kg = 8,000 g

c. 10 liters = _____ milliliters

d. 8 dm = _____ cm

e. 5 metric tons = _____ kg

SRB
215-217,
328

④ Find the volume of the prism.

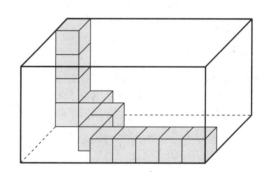

Volume = _____ cubic units

SRB
231-232

⑤ **Writing/Reasoning** Explain what happens to the value of the 8 as it moves one place to the left in Problem 2.

_____

_____

_____

SRB
66-67

Math Boxes

Practicing Multiplication Strategies

For Problems 1 and 2, solve one problem using U.S. traditional multiplication. Solve the other problem using any strategy. Show your work. Be sure to check that each answer makes sense.

SRB
44, 100,
102, 104

1
```
   23
*   6
____
```

2
```
   76
*   5
____
```

3 Choose Problem 1 or Problem 2. Explain how you checked to see whether your answer made sense.

_____

_____

_____

For Problems 4–7, do the following:

- Write a number model with a letter for the unknown.
- Solve the problem. Use U.S. traditional multiplication for at least one problem. Show your work.
- Write the answer.

4 Paula has 7 decks of cards. Each deck of cards has 52 cards in it. How many cards does she have in all?

Number model: _$52 * 7 = c$_

5 A bush is 21 inches tall. A tree is 5 times as tall as the bush. How tall is the tree?

Number model: _____

Answer: _____ cards

Answer: _____ inches

6 A fence has 45 sections. Each section is 6 meters long. How long is the fence?

Number model: _____

7 An apartment building has 9 apartments on each floor. There are 43 floors. How many apartments are in the building?

Number model: _____

Answer: _____ meters

Answer: _____ apartments

Math Boxes

---

① Solve.

a. $(3 \times 4) + (18 \div 3) =$ _____

b. $[7 \times (2 + 3)] - 20 =$ _____

c. $2 \times \{(81 \div 9) \div (9 \div 3)\} =$ _____

SRB
42-43

---

② Write a 5-digit number with

6 in the ones place,
3 in the thousands place,
1 in the hundreds place,
8 in the ten-thousands place,
and 0 in the tens place.

____ ____ , ____ ____ ____

SRB
66-67

---

③ Which expressions show 3,248 in expanded form?

Fill in the circle next to <u>all</u> that apply.

Ⓐ $32 \times 1,000 + 4 \times 10 + 8 \times 1$

Ⓑ $3 \,[1,000s] + 2 \,[100s] + 4 \,[10s] + 8 \,[1s]$

Ⓒ $3 \times 1,000 + 2 \times 100 + 4 \times 10 + 8 \times 1$

Ⓓ $3 \times 10^3 + 2 \times 10^2 + 4 \times 10^1 + 8 \times 10^0$

SRB
70

---

④ Find the volume of the prism.
Use the formula $V = l \times w \times h$.

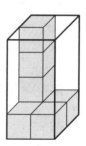

Volume = ____ × ____ × ____ = ____ units^3

SRB
231-233

---

⑤ Write in exponential notation.

a. $10 \times 10 \times 10 \times 10$ _____

b. $10 \times 10 \times 10$ _____

c. $10 \times 10 \times 10 \times 10 \times 10 \times 10 \times$ $10 \times 10 \times 10 \times 10 \times 10$ _____

SRB
68

---

⑥ Asher used 5 apples to make an apple pie. To make a jar of applesauce he needed twice as many apples as he needed for the pie plus two more. Write an expression that models how many apples Asher needed for the applesauce.

_____

SRB
42, 44

49

Practicing with Powers of 10

SRB
68-69

In our place-value system the powers of 10 are grouped into sets of three, which are called periods. We have periods for ones, thousands, millions, billions, and so on. When we write large numbers in standard notation, we separate these periods with commas. Mathematical language includes prefixes for the periods and other important powers of 10.

Periods										
	Millions			Thousands			Ones			
Billions	Hundred-millions	Ten-millions	Millions	Hundred-thousands	Ten-thousands	Thousands	Hundreds	Tens	Ones
10^9	10^8	10^7	10^6	10^5	10^4	10^3	10^2	10^1	10^0

Use the place-value chart and the prefixes chart to complete the following statements and fill in the missing exponents.

Prefixes	
tera-	trillion (10^{12})
giga-	billion (10^9)
mega-	million (10^6)
kilo-	thousand (10^3)
hecto-	hundred (10^2)
deca-	ten (10^1)
uni-	one (10^0)

1. The distance from Chicago to New Orleans is about 10^3, or one _____, miles.

2. A millionaire has at least $10^{\boxed{}}$ dollars.

3. The Moon is about 240,000, or _____ $* 10^{\boxed{}}$, miles from Earth.

4. A computer with a 1-terabyte hard drive can store approximately $10^{\boxed{}}$, or one _____, bytes of information.

5. The Sun is about $89 * 10^7$, or _____, miles from Saturn.

6. A 5-megapixel camera has a resolution of $5 * 10^{\boxed{}}$, or 5 _____ pixels.

7. What patterns do you notice in the following number sentences?

    $42 * 100 = 42 * 10^2 = 4{,}200$

    $42 * 1{,}000 = 42 * 10^3 = 42{,}000$

    $42 * 10{,}000 = 42 * 10^4 = 420{,}000$

_____

_____

Math Boxes

Math Boxes

1 Solve.

   a. 3 * 100 = _____

   b. 6 * 1,000 = _____

   c. 8 * 10,000 = _____

   d. 3 * _____ = 300,000

   e. 5 * _____ = 5,000,000

   SRB
   95-96

2 Write each number in exponential notation.

   a. 100 = _____

   b. 10,000 = _____

   c. 1,000,000 = _____

   d. 100,000 = _____

   e. 100,000,000 = _____

   SRB
   68

3 Find the volume of the prism.
Use the formula $V = B \times h$.

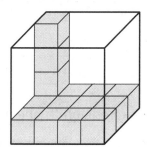

Area of the base = _____ units²

V = _____ × _____ = _____ units³

SRB
231-233

4 Make an estimate and then solve.

   a. _____   **b.** _____
        (estimate)   (estimate)

```
    3  4              5  2
 *     8           *     6
 _____          _____
```

SRB
83, 100,
102, 104

5 **Writing/Reasoning** Explain how you knew how many zeros belonged in the products in Problems 1a–1c.

_____

_____

_____

SRB
27-29,
95-96

Unit Conversion Number Stories

1 A dairy worker has 12 gallons of milk. She wants to pour it into 1-quart containers.

   a. How many quarts are in 1 gallon?

   _____ quarts

   b. How many quarts are in 12 gallons?

   _____ quarts

   c. How many quarts of milk does the dairy worker have?

   _____ quarts

   d. Write an expression to model the number story.

   _____

2 A seamstress sewed two pieces of fabric together. One piece was 3 feet long. The other was 8 inches long.

SRB
44, 215-
216, 328

   a. What is the length of the 3-foot piece of fabric in inches?

   _____ inches

   b. What is the total length of the new piece of fabric?

   _____ inches

   c. Write an expression to model both steps of the number story.

   _____

For Problems 3 and 4:

- Solve the problem.
- Write an expression to model the problem. Evaluate the expression to check your answer.

3 Two fifth-grade students had a running race. It took one student 2 minutes to run from one end of the playground to the other. It took the other student 98 seconds. How much faster was the second student's time?

Try This

4 A restaurant chef has 5 pounds of steak. He wants to cut it into 8-ounce portions to serve to customers. How many 8-ounce portions can he make?

Answer: _____ seconds

_____
(number model)

Answer: _____ portions

_____
(number model)

52

Math Boxes

Math Boxes

---

1 Complete.

a. $4 \times 3 =$ _____

b. $3 \times 10^3 =$ _____

c. $4 \times$ _____ $= 12,000$

d. $4 \times 3 \times$ _____ $= 12,000$

SRB 95-98

---

2 The figure below is a mathematical model of a blanket fort Sue built in her bedroom. Use the model to estimate the volume of the fort.

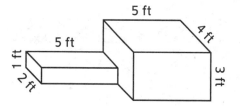

Volume: About _____ cubic feet

SRB 233-234

---

3 Make an estimate and solve.

$492 * 4 = ?$

_____
(estimate)

```
    4   9   2
*           4
_____
```

SRB 83, 100, 102, 104

---

4 Insert grouping symbols to make the number sentences true.

a. 5 * 4 – 2 = 10

b. 45 / 9 + 6 = 3

c. 2 * 4 ÷ 2 + 3 = 10

SRB 42-43

---

5 Which 6-digit numbers have
2 in the ones place,
7 in the thousands place, and
9 in the hundred-thousands place?

☐ 942,147   ☐ 749,124

☐ 947,142   ☐ 497,142

☐ 927,442

SRB 66-67

---

6 Fill in the missing digits.

```
        ☐
        7   6
*           4
_____
    ☐   ☐   4
```

SRB 102

53

Estimating and Multiplying with 2-Digit Numbers

Math Boxes

1 Complete. Use exponential notation for Parts d and e.

   a. $8 \times 10^2 =$ _____

   b. $3 \times 10^3 =$ _____

   c. $5 \times 10^4 =$ _____

   d. $2 \times$ _____ $= 2{,}000{,}000$

   e. $7 \times$ _____ $= 700{,}000$

SRB 95-96

2 Complete the table.

SRB 68

Standard Notation	Exponential Notation
10,000	
	10^5
	10^7
1,000,000	

3 Find the volume of the prism. Use the formula $V = B \times h$.

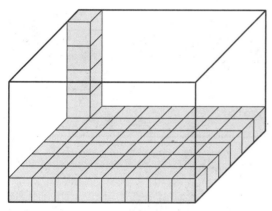

$V =$ _____ $\times$ _____ $=$ _____ units^3

SRB 231-233

4 Make an estimate and solve.

   a. _____   **b.** _____
     (estimate)   (estimate)

$$
\begin{array}{r}
6\ \ 4 \\
*\ \ \ \ \ 5 \\
\hline
\end{array}
\qquad
\begin{array}{r}
8\ \ 9 \\
*\ \ \ \ \ 3 \\
\hline
\end{array}
$$

SRB 83, 100, 102, 104

5 **Writing/Reasoning** Explain how you know the formula $V = B \times h$ tells you how many unit cubes could be packed into the prism in Problem 3.

_____

_____

_____

_____

SRB 233

Choosing Multiplication Strategies

Solve Problem 1 using U.S. traditional multiplication. Solve Problems 2–6 using any strategy. Show your work. Use your estimates to check whether your answers make sense. **SRB** 83. 100-104

1. $627 * 34 = ?$

   Estimate: _____

   $627 * 34 =$ _____

2. $148 * 8 = ?$

   Estimate: _____

   $148 * 8 =$ _____

3. $72 * 110 = ?$

   Estimate: _____

   $72 * 110 =$ _____

4. $436 * 65 = ?$

   Estimate: _____

   $436 * 65 =$ _____

5. A clerk ordered 72 boxes of paper clips. There are 250 paper clips in each box. How many paper clips are there in all?

   Estimate: _____

   Answer: _____ paper clips

6. A photo of a building is 18 centimeters tall. The real building is 892 times as tall. How tall is the building?

   Estimate: _____

   Answer: _____ centimeters

7. What strategy did you use to solve Problem 3? Explain why you chose that strategy.

   _____

   _____

Math Boxes

1 Complete.

   a. $7 \times 2 =$ _____

   b. $2 \times 10^2 =$ _____

   c. $7 \times 10^2 =$ _____

   d. $(7 \times 10^2) \times (2 \times 10^2) =$

   _____

   e. $700 \times 200 =$ _____

SRB
95-98

2 The figure below is a mathematical model of Remy's toy train car. Use the model to estimate the volume of the train car.

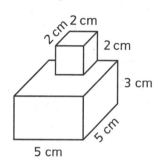

2 cm
2 cm
2 cm
2 cm
3 cm
5 cm
5 cm

Volume: About _____ cm³

SRB
233-234

3 Make an estimate and then solve.

   a. _____ **b.** _____
   (estimate)   (estimate)

$$\begin{array}{r} 2\;\;8\;\;7 \\ *\qquad 9 \\ \hline \end{array}$$

$$\begin{array}{r} 4\;\;1 \\ *\;1\;\;7 \\ \hline \end{array}$$

SRB
83, 100,
102, 104

4 Insert grouping symbols to make the number sentences true.

   a. $36 / 6 - 5 = 36$

   b. $25 * 8 - 3 = 125$

   c. $2 + 36 \div 12 - 10 = 19$

   d. $2 + 36 \div 12 - 6 = 8$

SRB
42-43

5 Write a 7-digit number with
5 in the hundred-thousands place,
2 in the tens place,
4 in the millions place,
6 in the ten-thousands place, and
0s in the other places.

_____

SRB
66-67

6 Fill in the missing digits.

$$\begin{array}{r} \boxed{} \\ 6\;\;2 \\ \times\qquad 8 \\ \hline \boxed{}\;\;9\;\;\boxed{} \end{array}$$

SRB
102

Invitations

Zoey is mailing invitations for a fifth-grade party. It takes her about 30 seconds to address 1 envelope.

1) About how many seconds would it take Zoey to address 10 envelopes? Show your work.

   About _____ seconds

2) About how many seconds would it take Zoey to address 100 envelopes? Show your work.

   About _____ seconds

58

Math Boxes

1 It took 32 minutes for Tara to walk to the store, 56 minutes to do her shopping, and 32 minutes to walk home. How many hours was Tara gone?

_____
(number model)

Answer: _____ hours

SRB
44, 216

2 Make an estimate and solve.

a. 3 1 2   _____
      ×   2 3   (estimate)

b. 4 9 6   _____
      ×   3 2   (estimate)

SRB
83, 100,
104

3 Which expression shows 5,892 in expanded form?
Fill in the circle next to the best answer.

○ **A.** $(5 \times 10^3) + (8 \times 10^2) + (9 \times 10^1) + (2 \times 10^0)$

○ **B.** $(5 \times 10^4) + (8 \times 10^3) + (9 \times 10^2) + (2 \times 10^1)$

○ **C.** $(5 \times 10^1) + (8 \times 10^2) + (9 \times 10^3) + (2 \times 10^4)$

SRB
70

4 Find the volume of the prism.
Use the formula $V = l \times w \times h$.

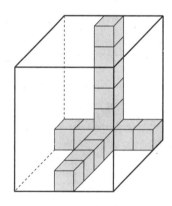

$V =$ _____ $\times$ _____ $\times$ _____ $=$ _____ units^3

SRB
231-233

5 **Writing/Reasoning** What method did you use to multiply in Problem 2a? Why did you choose that method?

_____

_____

_____

SRB
100-104

Using Multiples to Rename Dividends

Draw 2 cards to create a dividend. Draw one more card to create a divisor. Use multiples of the divisor to make equivalent names for the dividend. Then solve the division problem. Summarize your solution.

Example: Rosie drew 7, 4, and 3 and made her dividend 74 and her divisor 3. She listed multiples of 3 and completed the following name-collection box.

74
30 + 30 + 14
60 + 14
60 + 12 + 2

Multiples of 3: 3, 6, 9, 12, 15, 18, 21, 24, 27, 30...

60 + 12 + 2

60 / 3 = 20   12 / 3 = 4

74 / 3 → 24 R2

① Dividend: _____ Divisor: _____

Multiples of divisor: _____

_____

Summary: _____ / _____ → _____

② Dividend: _____ Divisor: _____

Multiples of divisor: _____

_____

Summary: _____ / _____ → _____

③ Dividend: _____ Divisor: _____

Multiples of divisor: _____

_____

Summary: _____ / _____ → _____

④ Dividend: _____ Divisor: _____

Multiples of divisor: _____

_____

Summary: _____ / _____ → _____

Practicing Unit Conversions

For Problems 1 and 2, complete the tables to show unit conversions.

1

Hours	Minutes
1	
2	
5	

2

Yards	Inches
1	
3	
10	

For Problems 3–6:

- Solve the problem.
- Write an expression to model the problem.
- Evaluate the expression to check your answer.

3 An adult African elephant can weigh up to 7 tons. At birth an African elephant weighs about 200 pounds. How many more pounds does an adult African elephant weigh than a newborn?

Answer: _____ pounds

_____
(number model)

4 A landscaper ordered 6 cubic yards of soil. So far she has used 90 cubic feet of soil. How many cubic feet of soil are left?

Hint: How many cubic feet are in 1 cubic yard?

Answer: _____ cubic feet

_____
(number model)

5 A rectangular room is 4 yards long and 5 yards wide. Lucas is covering the floor with tiles that are 1 square foot. How many tiles will he need?

Hint: Start by finding the area of the room in square yards.

Answer: _____ tiles

_____
(number model)

Try This

6 A football coach mixed 4 gallons of sports drink for his team. A serving of sports drink is 1 cup. How many servings of sports drink did the coach mix?

Answer: _____ servings

_____
(number model)

Math Boxes

1 Which model shows $\frac{5}{6}$ shaded?

Choose the best answer.

SRB
153, 155

2 Yasmin had 3 bananas. She shared them equally among the 4 people in her family. Write an expression that shows how much banana each person got.

_____

SRB
38, 44,
163-164

3 Marcus spent $\frac{1}{2}$ of his allowance on trading cards and $\frac{1}{4}$ of his allowance on snacks. Did he spend more on trading cards or snacks?

_____

SRB
174-175

4 Write two fractions equivalent to $\frac{1}{2}$.

_____   _____

Write two fractions equivalent to $\frac{1}{4}$.

_____   _____

SRB
165-166,
168, 170

5 Solve.

a. _____ $+ \frac{1}{8} + \frac{1}{8} = \frac{3}{8}$

b. $\frac{1}{4} +$ _____ $+ \frac{1}{4} = \frac{3}{4}$

c. $\frac{1}{5} +$ _____ $= \frac{4}{5}$

d. $\frac{2}{9} + \frac{5}{9} =$ _____

SRB
186

6 Brielle is buying yarn to knit a scarf. She needs to know the area of the scarf she will knit to choose the right package of yarn. What is the area of a scarf that is 4 feet long and $\frac{1}{2}$ foot wide?

4 ft

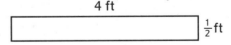
$\frac{1}{2}$ ft

Area = _____ square feet

SRB
186, 225

Partial-Quotients Division

For Problems 1–4, make an estimate. Then use partial-quotients division to solve. Show your work on the computation grid.

SRB
84,
108-112

1 234 / 11 → ?

Estimate: _____

Answer: _____

2 825 / 15 → ?

Estimate: _____

Answer: _____

3 3,518 / 30 → ?

Estimate: _____

Answer: _____

4 6,048 / 54 → ?

Estimate: _____

Answer: _____

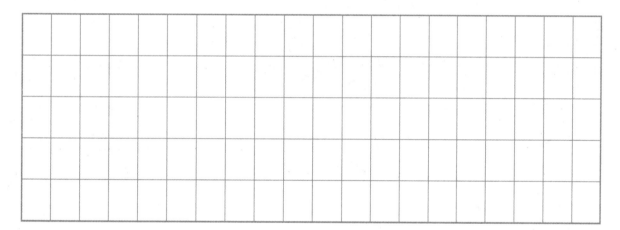

Try This

5 Complete the area model on the right to show your solution for Problem 2.

Hint: Think of Problem 2 as: *If the area of a room is 825 square feet and the length of the room is 15 feet, how wide is the room?*

Area (Dividend): _____

Length (Divisor): _____

Width (Quotient): _____

Math Boxes

1 Solve.

a. 45 / 9 = _____

b. 450 / 9 = _____

c. 4,500 / 9 = _____

d. 32 / 8 = _____

e. 320 / 8 = _____

f. 3,200 / 8 = _____

SRB
106

2 Find the volume of the prism.
Use the formula $V = B \times h$.

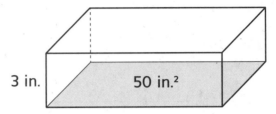

3 in.   50 in.²

V = _____ × _____ = _____ in.³

SRB
233

3 True or false?

In the number 23,916:

a. the digit 3 is worth 3,000.
   ○ true   ○ false

b. the digit 9 is worth 90.
   ○ true   ○ false

c. the digit 2 is worth 20,000.
   ○ true   ○ false

d. the digit 1 is worth 100.
   ○ true   ○ false

SRB
66-67

4 Fill in the missing digits.

a.
```
        4    □
     2 . 8   2
  ×          6
  ─────────────
  □ ,  6  □   2
```

b.
```
       □    3
       3  8  6
  ×          5
  ─────────────
  1,  □  3  □
```

SRB
102

5 **Writing/Reasoning** Explain how you solved Problem 1e.

_____

_____

_____

SRB
106

Partial Quotients with Multiples

For Problems 1–4, make an estimate. Then use partial-quotients division to solve. Show your work. You can make lists of multiples on *Math Masters,* page TA10 to help you.

SRB
72, 84,
108-110

(1) 1,647 / 28 → ?

Estimate: _____

Answer: _____

(2) 4,319 / 42 → ?

Estimate: _____

Answer: _____

(3) 2,628 / 36 → ?

Estimate: _____

Answer: _____

(4) 9,236 / 41 → ?

Estimate: _____

Answer: _____

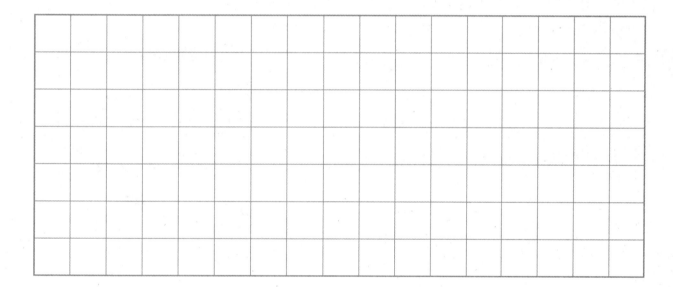

Try This

(5) Paul drew the area model at the right for his solution to Problem 1. What partial quotients did he use to solve the problem?

_____

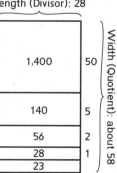

Area (Dividend): 1,647
Length (Divisor): 28

Width (Quotient): about 58

1,400	50
140	5
56	2
28	1
23	

Math Boxes

Math Boxes

---

① Yao bought 3 feet of blue ribbon and 24 inches of red ribbon. The ribbon costs $2.00 per foot. What is the total cost of the ribbon?

_____
(number model)

Answer: _____ dollars

② Fill in the circle next to the best estimate for the problem below. Then solve.

Estimate:

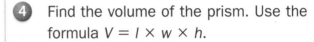
$$\begin{array}{r} 2\ \ 1\ \ 7 \\ \times\ \ 1\ \ 9\ \ 8 \\ \hline \end{array}$$

Ⓐ 40,000

Ⓑ 4,000

Ⓒ 20,000

Ⓓ 400,000

SRB
83,
100-104

SRB
44, 216

---

③ Write 72,658 in expanded form using exponents to write powers of 10.

_____

_____

SRB
70

④ Find the volume of the prism. Use the formula $V = l \times w \times h$.

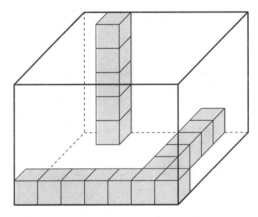

$V =$ ____ × ____ × ____ = _____ units^3

SRB
231-233

---

⑤ **Writing/Reasoning** Explain how you solved Problem 1.

_____

_____

_____

_____

SRB
44, 216

1 Solve.

a. 42 / 6 = _____

b. 420 / 6 = _____

c. 4,200 / 60 = _____

d. 81 / 9 = _____

e. 81,000 / 90 = _____

SRB
106

2 Find the volume of the prism.
Use the formula $V = B \times h$.

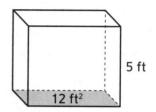

5 ft

12 ft²

$V =$ _____ $\times$ _____ $=$ _____ ft³

SRB
233

3 Write the value of the **boldface digit** in each number.

a. 3**9**0 _____

b. **8**,092 _____

c. 3**5**,047 _____

d. 2**3**2,591 _____

e. **4**97,214 _____

SRB
66-67

4 Fill in the missing digits.

a.
```
        □   □
    4   5   3
×               4
  1,  □   □   2
```

b.
```
          □
    3   2   7
×               3
    9   □   □
```

SRB
102

5 **Writing/Reasoning** In Problem 3, how would the value of the boldface digits change if they moved one place to the right?

_____

_____

_____

SRB
66-67

Interpreting Remainders

For each problem:

- Create a mathematical model.
- Solve the problem. Show your work.
- Tell what the remainder represents.
- Decide what to do with the remainder. Explain what you did.

SRB
12-14,
113

1. Basketballs are on sale for $12, including tax. How many basketballs can the gym teacher buy with $40?

   Mathematical model:

   Quotient: _____ Remainder: _____

   What does the remainder represent?

   _____

   _____

   Answer: The gym teacher can buy

   _____ basketballs.

   Circle what you did with the remainder.

   Ignored it

   Rounded the quotient up

   Why?

   _____

   _____

   _____

2. You are organizing a trip to a museum for 110 students, teachers, and parents. If each bus can seat 25 people, how many buses do you need?

   Mathematical model:

   Quotient: _____ Remainder: _____

   What does the remainder represent?

   _____

   _____

   Answer: I need _____ buses.

   Circle what you did with the remainder.

   Ignored it

   Rounded the quotient up

   Why?

   _____

   _____

   _____

Interpreting Remainders
(continued)

SRB
12-14,
113

3 Mrs. Maxwell has 60 pens to pass out to her class. There are 27 students in her class. How many pens will each student get if everyone is given a fair share?

Mathematical model:

Quotient: _____ Remainder: _____

What does the remainder represent?

_____

_____

Answer: Each student will get

_____ pens.

Circle what you did with the remainder.

   Ignored it

   Rounded the quotient up

Why?

_____

_____

_____

4 Choose one of your mathematical models. Explain how it helped you solve the problem.

_____

_____

_____

_____

_____

_____

_____

_____

Math Boxes

1 Match the letters to the fractions they represent on the number line.

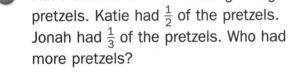

0   a   b   c   1

$\frac{1}{2}$ = _____

$\frac{1}{4}$ = _____

$\frac{3}{4}$ = _____

SRB
153, 158,
161

2 Ben has 3 cans of food to feed his cat for 5 days. Write an expression that shows how much of a can Ben should feed his cat each day.

_____

SRB
38, 44,
163 164

3 Katie and Jonah were sharing a bag of pretzels. Katie had $\frac{1}{2}$ of the pretzels. Jonah had $\frac{1}{3}$ of the pretzels. Who had more pretzels?

_____

SRB
174-175

4 Write two fractions equivalent to $\frac{1}{3}$.

_____ _____

Write two fractions equivalent to $\frac{2}{3}$.

_____ _____

SRB
165-166,
168, 170

5 Solve.

a. $\frac{1}{7} + \frac{1}{7} + \frac{1}{7}$ = _____

b. $\frac{1}{12} +$ _____ $= \frac{6}{12}$

c. $\frac{3}{6} + \frac{1}{6} +$ _____ $= \frac{5}{6}$

d. _____ $+ \frac{2}{9} = \frac{5}{9}$

e. $\frac{1}{4} + \frac{1}{4} + \frac{1}{4}$ = _____

SRB
186

6 Ricardo wants to cover a shelf with shelf liner. The shelf is 4 feet wide and $\frac{2}{3}$ feet deep. What is the area of the shelf?

Area = _____ square feet

SRB
12-14,
186, 225

Solving Fair Share
Number Stories

Use fraction circle pieces or a drawing to model each number story. Then solve.

SRB
163-164

1 Mary and her two friends were working on a science project. They shared 1 pizza equally as a snack. How much pizza did each person get?

Models:

Solution: _____

2 Jose is taking care of a neighbor's cat. The neighbor will be gone for 5 days and left 3 cans of cat food. The cat is supposed to eat the same amount each day. How much food should Jose give the cat each day?

Solution: _____

3 A school received a shipment of 4 boxes of paper. The school wants to split the paper equally among its 3 printers. How much paper should go to each printer?

Solution: _____

4 Adrian brought 2 loaves of olive bread to school for a class celebration. There were 12 people who wanted to try the bread. They decided to split the loaves evenly. How much bread did each person receive?

Solution: _____

Multiplication and Division

For Problems 1–3, make an estimate. Then solve using U.S. traditional multiplication. Use your **SRB** 83, 103, 108-112 estimates to check whether your answers make sense.

① _____
(estimate)

② _____
(estimate)

③ _____
(estimate)

```
    28
  * 57
```

```
   643
  * 81
```

```
   706
 * 145
```

For Problems 4 and 5, make an estimate. Then solve using partial-quotients division. Use your estimates to check whether your answers make sense.

④ _____
(estimate)

⑤ _____
(estimate)

⑥ Complete the area model to represent your solution to Problem 4.

$32)\overline{4,168}$

$56)\overline{7,211}$

Area (Dividend): _____

Length (Divisor): _____

Width (Quotient): _____

Answer: _____ R_____

Answer: _____ R_____

72

Math Boxes

1 Insert grouping symbols to make true number sentences.

   a. 19 + 41 * 3 = 180

   b. 5 = 16 / 2 + 2 − 5

   c. 24 ÷ 8 + 4 * 3 = 6

   d. 24 ÷ 8 + 4 * 3 = 15

   e. 1 = 16 / 2 + 2 − 3

SRB
42-43

2 Complete the following equivalents.

   a. 1 pint = _____ cups

   b. 6 quarts = _____ pints

   c. 1 quart = _____ cups

   d. 3 gallons = _____ quarts

   e. 1 gallon = _____ cups

SRB
215-217,
328

3 Write in standard notation or expanded form.

   a. 82,913 = _____

   _____

   b. $2 \times 100{,}000 + 6 \times 10{,}000 + 1 \times 1{,}000 + 9 \times 100 + 4 \times 10 + 5 \times 1 =$ _____

   c. $5 \times 10^3 + 2 \times 10^2 + 0 \times 10^1 + 7 \times 10^0 =$ _____

SRB
70

4 What is the volume of the prism?

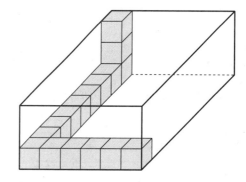

Volume = _____ cubic units

SRB
231-233

5 Solve mentally by breaking the dividend into smaller parts that are easier to divide. Write the equivalent name you used.

96 ÷ 3 → _____

Equivalent name for 96:

_____

SRB
107

6 Solve. Use an estimate to check whether your answer makes sense.

      7 2 8
   × 1 2
   ‾‾‾‾‾

_____
(estimate)

SRB
83,
100-104

73

Writing Division Number Stories

Record the fractions you are assigned. For each fraction, write a division number sentence with the fraction as the quotient. Then write a number story to match the number sentence. SRB 163-164

1. Fraction: _____

   Division number sentence: _____

   Number story: _____

   _____

   _____

   _____

2. Fraction: _____

   Division number sentence: _____

   Number story: _____

   _____

   _____

   _____

3. Fraction: _____

   Division number sentence: _____

   Number story: _____

   _____

   _____

   _____

More Practice with Fair Shares

Solve each number story. You can use fraction circle pieces or drawings to help. Write a number model to show how you solved each problem.

SRB
163-164

1. Davita brought 6 granola bars for herself and the 7 other girls in her camp group for a snack. If they share them equally, what fraction of a granola bar will each girl get?

   Solution: _____ granola bar

   Number model: _____

2. Lucas is making 12 jumbo muffins to sell at his class bake sale. He has 2 bowls full of batter. What fraction of a bowl of batter should Lucas put in each muffin cup?

   Solution: _____ bowl

   Number model: _____

3. Ms. Cox is combining bottles of hand sanitizer. She has 11 small bottles of sanitizer she wants to divide equally among 3 large containers. How many small bottles should she empty into each large container?

   Solution: _____ small bottles

   Number model: _____

4. Write a division number story with an answer of $\frac{12}{8}$.

   _____

   _____

   _____

   _____

   Number model: _____

75

Math Boxes

1 Solve.

  a. $4 \times 100 =$ _____

  b. $4 \times 10^2 =$ _____

  c. $6 \times 10^3 =$ _____

  d. $6 \times 1,000 =$ _____

SRB
95-96

2 Solve.

$25\overline{)578}$

$578 \div 25 \rightarrow$ _____

SRB
108-110

3 Find the area of a table top that is $3\frac{1}{3}$ feet by 2 feet.

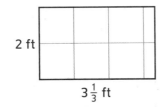

2 ft

$3\frac{1}{3}$ ft

Area = _____ square feet

_____
(number model)

SRB
224-225

4 Kayin buys 6 envelopes for 35 cents each and 6 stamps for 48 cents each. Which expression models how much money Kayin spends?

Fill in the circle next to the best answer.

○ **A.** $(6 + 6) * (35 + 48)$

○ **B.** $(6 * 6) + (35 + 48)$

○ **C.** $(6 * 35) + (6 * 48)$

SRB
42, 44

5 **Writing/Reasoning** Describe a pattern you noticed in Problem 1.

_____

_____

_____

SRB
95-96

Division Number Stories with Remainders

For each number story:

• Write a number model with a letter for the unknown.
• Solve. Show your work in the space provided. You may draw a picture to help.
• Decide what to do with the remainder. Explain what you did and why.

SRB
44.
113-114

---

1 Rebecca and her two sisters made pancakes for breakfast. They made 16 pancakes for 5 people. They want to make sure each person gets an equal serving. How many pancakes will each person get?

Number model: _____

Quotient: _____ Remainder: _____

Answer: Each person will get _____ pancakes.

Circle what you did with the remainder.

　Ignored it

　Reported it as a fraction

　Rounded the quotient up

Why? _____

_____

_____

2 Louis's soccer team is taking a bus to a tournament. They have 32 reusable water bottles. Their water carriers hold 6 bottles each. How many carriers will Louis's team need to bring all of their water bottles on the bus?

Number model: _____

Quotient: _____ Remainder: _____

Answer: Louis's team needs _____ carriers.

Circle what you did with the remainder.

　Ignored it

　Reported it as a fraction

　Rounded the quotient up

Why? _____

_____

_____

77

Division Number Stories with Remainders (continued)

Lesson 3-3

DATE   TIME

3 Mariana saved $80 from her babysitting job. She wants to buy some shirts and pants that are on sale at her favorite store for $17 each. How many items of clothing can she buy?

Number model: _____

Quotient: _____ Remainder: _____

Answer: Mariana can buy

_____ items.

Circle what you did with the remainder.

Ignored it

Reported it as a fraction

Rounded the quotient up

Why? _____

_____

_____

4 Jeremy wants to read 100 more books by the end of the school year. There are 36 weeks of school. How many books does Jeremy need to read each week?

SRB
44, 109, 113-114

Number model: _____

Quotient: _____ Remainder: _____

Answer: Jeremy needs to read

_____ books each week.

Circle what you did with the remainder.

Ignored it

Reported it as a fraction

Rounded the quotient up

Why? _____

_____

_____

78

Math Boxes

Math Boxes

---

1 Insert grouping symbols to make true number sentences.

a. 4 * 8 − 5 = 12

b. 2 + 7 * 7 = 51

c. 91 / 4 − 3 = 91

d. 20 * 2 + 1 + 3 / 9 = 7

e. 60 + 12 / 30 + 6 = 2

SRB
42-43

---

2 Complete the following equivalents.

a. 1 cup = _____ ounces

b. 1 pint = _____ ounces

c. 1 quart = _____ ounces

d. 4 quarts = _____ ounces

e. 1 gallon = _____ ounces

SRB
215-217,
328

---

3 Which shows 672,891 in expanded form?

Fill in the circle next to the best answer.

(A) $6 \times 10{,}000 + 7 \times 1{,}000 + 2 \times 100 + 8 \times 10 + 9 \times 1 + 1 \times 0$

(B) 6 [100,000s] + 7 [1,000s] + 2 [100s] + 8 [10s] + 9 [1s] + 1

(C) $6 \times 10^5 + 7 \times 10^4 + 2 \times 10^3 + 8 \times 10^2 + 9 \times 10^1 + 1 \times 10^0$

(D) 670,000 + 2,800 + 90 + 1

SRB
70

---

4 What is the volume of the prism?

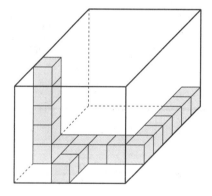

Volume = _____ units^3

SRB
231-233

---

5 Solve mentally by breaking the dividend into smaller parts that are easier to divide. Write the equivalent name you used.

840 ÷ 20 → _____

Equivalent name for 840:

_____

SRB
107

---

6 Solve. Use an estimate to check your answer.

```
  1,  1  1  3
×        3  7
_____
```

_____
(estimate)

SRB
83,
100-104

79

Fractions on a Number Line

① Use the number line below for the Math Message.

SRB
158-161,
171-173

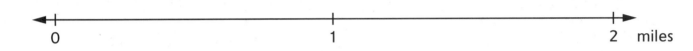

0   1   2   miles

② Gary ran $1\frac{2}{3}$ miles and Lena ran $\frac{7}{6}$ mile. Who ran farther? Use the number lines below to help you answer the question.

a. Divide this number line to show thirds. Label each tick mark. Then place a dot at $1\frac{2}{3}$.

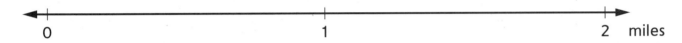

0   1   2   miles

b. Divide this number line to show sixths. Label each tick mark. Place a dot at $\frac{7}{6}$.

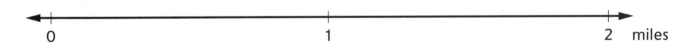

0   1   2   miles

c. Who ran farther? _____

③ Which number is greater? Circle the greater number in each pair. Use the Fraction Number Lines Poster or fraction circle pieces to help you.

a. $\frac{5}{8}$ or $\frac{9}{10}$ b. $\frac{5}{3}$ or $1\frac{5}{6}$ c. $2\frac{1}{4}$ or $\frac{20}{12}$ d. $\frac{9}{6}$ or $\frac{13}{12}$

80

Fractions on a Number Line

(continued)

4 Rachel and Dan are growing plants in science class. Rachel reports that her plant is $1\frac{1}{4}$ inches tall. Dan says his plant is $\frac{5}{2}$ inches tall.

   a. Whose plant is taller? _____

   b. How do you know? _____

   _____

   _____

For Problems 5–10, rename each fraction as a mixed number or each mixed number as a fraction greater than 1. You may use the Fraction Number Lines Poster, fraction circle pieces, or division.

5 $\frac{5}{3} = $ _____

6 $1\frac{7}{9} = $ _____

7 $\frac{11}{8} = $ _____

8 $1\frac{5}{6} = $ _____

9 $\frac{16}{5} = $ _____

10 $2\frac{1}{3} = $ _____

Try This

11   **a.** Rename $\frac{34}{8}$ as a mixed number. _____

   b. Explain your reasoning.

   _____

   _____

   _____

Math Boxes

Math Boxes

(1) Solve.

    a. $3 * 10^1 =$ _____

    b. $3 * 10^2 =$ _____

    c. $3 * 10^3 =$ _____

    d. $3 * 10^4 =$ _____

Write a number sentence that follows the pattern above.

_____ * _____ = _____

SRB
95-96

(2) Solve.

$$32\overline{)6{,}572}$$

$6{,}572 \div 32 \rightarrow$ _____

SRB
108-110

(3) What is the area of a garden plot that measures $6\frac{1}{2}$ feet by 4 feet?

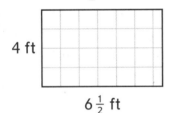

4 ft

$6\frac{1}{2}$ ft

Area = _____ square feet

_____
(number model)

SRB
224-225

(4) Jamar bought eight 6-packs of juice boxes for his family. His grandmother bought 3 more 6-packs. Write an expression that models how many juice boxes they bought.

_____

SRB
44

(5) **Writing and Reasoning** Ari made a sketch to solve Problem 2. Use Ari's sketch to explain how you think he solved the problem.

Area: 6,572

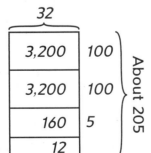

32

3,200	100
3,200	100
160	5
12	

About 205

_____

_____

_____

_____

_____

SRB
8-9,
111-112

82

Division Top-It

Harjit is playing a version of *Division Top-It* with a friend. In this version each player turns over 3 number cards and places them as the digits in the division problem below. The player with the larger quotient wins the round.

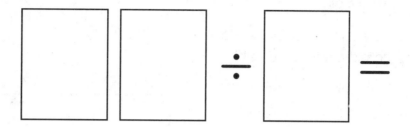

Harjit turns over a 6, a 3, and a 9. How do you think Harjit should place her cards to get the largest possible quotient? Explain your thinking.

Math Boxes

1 Which number stories have the answer $\frac{3}{4}$? Circle ALL that apply.

   A. Four dogs shared 3 dog treats. How many treats did each dog get?

   B. Three friends shared 4 oranges. How much orange did each friend eat?

   C. Four dogs each ate 3 dog treats. How many treats did the dogs eat?

   D. Four friends shared 3 oranges. How much orange did each friend eat?

SRB
163-164

2 Estimate and fill in the missing digits.

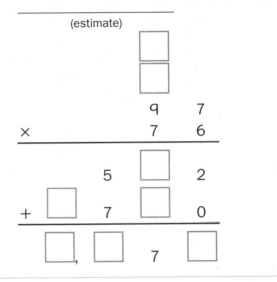

_____
(estimate)

SRB
83, 103

3 Complete the table.

Exponential Notation	Standard Notation
10^0	
10^1	
	100
	100,000
	10,000,000

SRB
68

4 Solve.

   a. 63 / 9 = _____

   b. 630 / 9 = _____

   c. 630 / 90 = _____

   d. 6,300 / 900 = _____

   e. 63,000 / 9 = _____

SRB
106

5 Find the volume of a box with a base area of 24 in.² and a height of 12 in. Use the formula $V = B \times h$.

Volume = _____ × _____

Volume = _____ in.³

SRB
233

6 Place the fractions on the number line.

$\frac{1}{2}$ $\frac{3}{4}$ $\frac{1}{4}$

0   1

_____ _____ _____

SRB
158, 161

Checking for Reasonable Answers

SRB
10-11,
181-185

Christopher solved some fraction problems. Do Christopher's answers make sense?
Circle Yes or No. Then write an argument to show how you know.

Name _Christopher_

Solve.

1 $\frac{2}{7} + \frac{1}{2} = \frac{3}{9}$	**Conjecture:** Does answer 1 make sense? **Yes No** **Argument:** _____ _____ _____
2 Write >, <, or =. $\frac{9}{10} \underline{\quad > \quad} \frac{7}{8}$	**Conjecture:** Does answer 2 make sense? **Yes No** **Argument:** _____ _____ _____
3 $\frac{7}{12} + \frac{1}{4} = \frac{8}{12}$	**Conjecture:** Does answer 3 make sense? **Yes No** **Argument:** _____ _____ _____
4 $\frac{8}{9} + \frac{1}{3} = \frac{8}{12}$	**Conjecture:** Does answer 4 make sense? **Yes No** **Argument:** _____ _____ _____

Interpreting Remainders in Number Stories

For each number story, write a number model with a letter for the unknown. Then solve. Show your work in the space provided. You can draw a picture to help. Decide what to do with the remainder and explain what you did.

SRB
44, 108,
113-114

1 A cook has 250 ounces of cheese for 80 individual pizzas.
Each pizza gets the same amount of cheese.
How much cheese should the cook put on each pizza?

Number model: _____

Quotient: _____ Remainder: _____

Answer: The cook should put _____ ounces of cheese
on each pizza.

Circle what you did with the remainder.

Ignored it   Reported it as a fraction   Rounded the quotient up

Why? _____

_____

_____

2 35 people are attending a game night. Each table seats
4 people. How many tables are needed?

Number model: _____

Quotient: _____ Remainder: _____

Answer: _____ tables are needed.

Circle what you did with the remainder.

Ignored it   Reported it as a fraction   Rounded the quotient up

Why? _____

_____

_____

Math Boxes

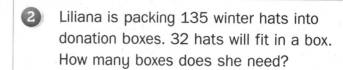

---

① Liz bought four 2-quart packages of strawberries. How many gallons of strawberries did she buy?

_____
(number model)

Answer: _____ gallons

SRB
44, 214-
217, 328

② Liliana is packing 135 winter hats into donation boxes. 32 hats will fit in a box. How many boxes does she need?

_____
(number model)

Quotient: _____ Remainder: _____

Answer: She needs _____ boxes.

SRB
44, 109,
113

---

③ Four friends equally shared 6 cups of soup. How much soup did each friend get? Circle ALL that apply.

A. $\frac{6}{4}$ cups

B. $1\frac{2}{4}$ cups

C. $1\frac{1}{2}$ cups

D. $\frac{4}{6}$ cup

SRB
163-164,
170-171

④ Complete.

a. Write a number in which a 5 is worth 500. _____

Write a number in which a 5 is worth 10 times as much. _____

b. Write a number in which a 7 is worth 70,000. _____

Write a number in which a 7 is worth $\frac{1}{10}$ as much. _____

SRB
66-67

---

⑤ **Writing/Reasoning** What did you decide to do with the remainder in

Problem 2? Why? _____

_____

_____

SRB
113

87

Using Benchmarks to Make Estimates

SRB
182

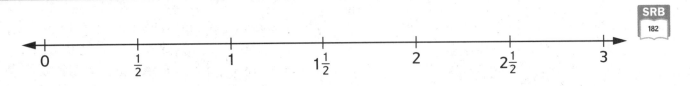

Estimate the sum or difference for each fraction number story. Place an X on the number line to represent your estimate. Then circle the best answer.

1. Micah bought $1\frac{1}{3}$ pounds of grapes and $1\frac{1}{2}$ pounds of bananas. About how many pounds of fruit did Micah buy?

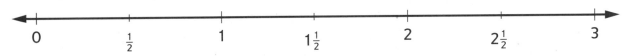

How much fruit?   about 2 pounds   about $2\frac{1}{2}$ pounds   about 3 pounds

Explain your thinking.

_____

_____

2. Chloe has $2\frac{1}{2}$ yards of fabric. She will use about $\frac{3}{8}$ yard to make a scarf. How many yards of fabric will she have left?

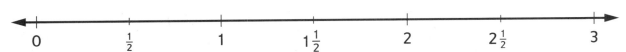

How much fabric is left?   about $1\frac{1}{2}$ yards   about 2 yards   about 3 yards

Explain your thinking.

_____

_____

Try This

3. The perimeter of a triangle is 10 inches. One side is $3\frac{9}{16}$ inches long. Another side is $4\frac{5}{8}$ inches long. About how many inches long is the third side? Explain how you estimated.

_____

_____

_____

Math Boxes

1 Write a division number story with an answer of $\frac{3}{5}$.

_____

_____

_____

_____

SRB
163-164

2 Estimate. Fill in the missing digits.

Estimate: _____

$$
\begin{array}{r}
\boxed{} \\
\boxed{} \\
4 \quad 1 \quad 4 \\
\times \qquad 7 \quad 7 \\
\hline
\boxed{}\ 8\ \boxed{}\ 8 \\
+\ 2\ \ 8\ \boxed{}\ \boxed{}\ 0 \\
\hline
\boxed{}\boxed{},\ 8\ \ 7\ \boxed{}
\end{array}
$$

SRB
83, 103

3 Rewrite each number in standard or exponential notation.

a. $10^3 =$ _____

b. $10,000 =$ _____

c. $10^5 =$ _____

d. $1,000,000 =$ _____

e. $10^8 =$ _____

SRB
68

4 Solve.

How many 8s in 72? _____

How many 800s in 72,000? _____

How many 5s in 450,000? _____

How many 3,000s in 270,000? _____

How many 90s in 63,000? _____

SRB
106

5 Find the volume of a shipping container that is 20 feet long, 8 feet wide, and 8 feet tall. Use the formula $V = l \times w \times h$.

$V =$ _____ $\times$ _____ $\times$ _____

Volume = _____ ft^3

SRB
233

6 Place the fractions on the number line.

$\frac{1}{3}$ $\qquad$ $\frac{2}{3}$ $\qquad$ $\frac{4}{3}$ $\qquad$ $\frac{5}{3}$

0   1   2

SRB
158-161

89

Renaming Fractions and Mixed Numbers

Solve by following the steps. You can use fraction circles, number lines, or drawings to help.

SRB
171-173

1. Find another name for $2\frac{3}{4}$.

   - Show $2\frac{3}{4}$.

   - Break apart 1 whole into $\frac{4}{4}$.

   Name: _____

2. Find another name for $\frac{14}{3}$.

   - Show $\frac{14}{3}$.

   - Make as many groups of 3 thirds as you can.

   - Trade each $\frac{3}{3}$ for 1 whole.

   Name: _____

Write another name for each mixed number that has the same denominator. Check that your trades are fair and record them.

Example: $3\frac{8}{6}$

Name: _____ $2\frac{14}{6}$ _____

Trade: _____ *1 whole for $\frac{6}{6}$* _____

3. $2\frac{4}{5}$

   Name: _____

   Trade: _____

4. $1\frac{12}{10}$

   Name: _____

   Trade: _____

Fill in the missing whole number or missing numerator.

5. $1\frac{4}{3} = 2\frac{\boxed{}}{3}$

6. $\frac{\boxed{}}{5} = 4\frac{2}{5}$

7. $2\frac{9}{2} = 4\frac{\boxed{}}{2}$

8.
   a. Mojo the monkey has 2 whole bananas and 5 half-bananas. Write a mixed number to show how many bananas Mojo has. _____ bananas

   b. Manny the monkey has 4 whole bananas and 1 half-banana. Do Mojo and Manny have the same amount of banana? Explain how you know.

   _____

   _____

   c. Marcus the monkey has the same amount of banana as Mojo. He only has half-bananas. How many half-bananas does he have? Explain your answer.

   _____

   _____

90

Connecting Fractions and Division

Write a division expression to model each story, then solve. You can use fraction circles or draw pictures to help.

SRB
27-29,
163-164

1. Olivia is running a 3-mile relay race with 3 friends. If the 4 of them each run the same distance, how many miles will each person run?

   Division number model: _____

   Fractional answer: _____

2. Chris has 3 pints of blueberries for fruit salad. If he splits the blueberries equally into 8 serving bowls, how many pints of blueberries will be in each bowl?

   Division number model: _____

   Fractional answer: _____

3. Four students shared 9 packages of pencils equally. How many packages of pencils did each student get?

   Division number model: _____

   Fractional answer: _____

Use your answers to Problems 1–3 to answer the questions below.

4. Compare the numbers in each division number model with your fractional answer. What do you notice?

   _____

   _____

   _____

5. Write a rule for finding the fractional answer to a division problem by using the dividend and divisor.

   _____

   _____

6. Write and solve your own number story using your rule.

   _____

   _____

Math Boxes

1 Noelle bought two $\frac{1}{2}$-gallon cartons of lemonade for her lemonade stand. How many ounces of lemonade did she buy?

_____
(number model)

Answer: _____ ounces

SRB
214-217,
328

2 The student council has $289 to spend on decorations for the fall festival. If decorations for each table cost $13, how many tables can they decorate?

_____
(number model)

Quotient: _____ Remainder: _____

Answer: They can decorate _____ tables.

SRB
44, 109,
113

3 The fifth-grade teachers ordered 3 boxes of clay. They wanted to share it equally among their four classes. How much clay does each class get?

Division number model:

_____

Answer: _____ box of clay

SRB
163-164

4 Move the digits in 625,134 to create a new number.

Move the 2 so it is worth $\frac{1}{10}$ as much.

Move the 3 so it is worth 10 times as much.

Move the 5 so it is worth 50,000.

Move the 4 so its value changes to $4 \times 100,000$.

Move the 1 and the 6 so that the sum of their values is 16.

Write the new number:

_____

SRB
66-67

5 **Writing/Reasoning** Draw a picture to show how you solved Problem 3.

SRB
12-14,
163-164

Addition and Subtraction Number Stories

For each story:

- Write a number model with a letter for the unknown.
- Make an estimate.
- Solve. You can use fraction circle pieces, a drawing, or a number line to help.
- Use your estimate to check whether your answer makes sense.

SRB
178-185

1. Andrea had $1\frac{1}{5}$ liters of water. She drank $\frac{3}{5}$ liter. How much did she have left?

    Number model: _____

    Estimate: _____

    Answer: _____ liter

2. A table is $2\frac{8}{12}$ feet tall and a lamp on it is $1\frac{5}{12}$ feet tall. What is their total height?

    Number model: _____

    Estimate: _____

    Answer: _____ feet

3. A chef had $2\frac{5}{8}$ pitas. She used $1\frac{7}{8}$ pitas. How many pitas does she have left?

    Number model: _____

    Estimate: _____

    Answer: _____ pita

4. Niko rode a bike $2\frac{3}{10}$ miles. Then he rode another $2\frac{8}{10}$ miles. How far did he ride?

    Number model: _____

    Estimate: _____

    Answer: _____ miles

5. Explain how you solved Problem 3.

    _____

    _____

    _____

1 Allison was making pancakes. She needed $\frac{1}{4}$ cup of vegetable oil and $\frac{3}{4}$ cup of milk. How many cups of liquid did she need?

_____
(number model)

Answer: _____ cup(s) of liquid

SRB
178-180, 186

2 Solve.

a.
```
  2 2 5
×     2
_____
```

b.
```
  2 2 5
×     4
_____
```

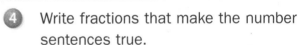
SRB
100, 102, 104

3 Solve.

a. $(4 * 12) + 8 =$ _____

b. _____ $= (32 / 16) / 2$

c. $(65 + 83) / (3 - 1) =$ _____

d. _____ $= 3 + \{32 / (16 / 2)\}$

SRB
42-43

4 Write fractions that make the number sentences true.

a. _____ $+$ _____ < 1

b. _____ $-$ _____ < 1

c. _____ $+$ _____ > 2

d. _____ $+$ _____ $< 1\frac{1}{2}$

SRB
181-184

5 **Writing/Reasoning** Morton says he can compare the products in Problem 2 without multiplying. Explain how.

_____

_____

_____

_____

SRB
46

Adding Fractions with Circle Pieces

Make an estimate. Then use your fraction circle pieces to find the sum. Use the red circle as the whole. Remember to think about using same-size pieces.

SRB 155, 189

Write a number sentence to show how you used equivalent fractions to find the sum.

Example: $\frac{1}{2} + \frac{1}{8} = ?$

Estimate: *Between $\frac{1}{2}$ and 1*

Show $\frac{1}{2} + \frac{1}{8}$ with fraction pieces.

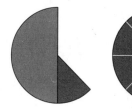

Cover the $\frac{1}{2}$ piece with four $\frac{1}{8}$ pieces to show that $\frac{1}{2}$ is the same as $\frac{4}{8}$.

Sum: $\frac{5}{8}$

Number sentence: $\frac{4}{8} + \frac{1}{8} = \frac{5}{8}$

1 $\frac{2}{3} + \frac{1}{6} = ?$

Estimate: _____

Sum: _____

Number sentence: _____

2 $\frac{2}{5} + \frac{3}{10} = ?$

Estimate: _____

Sum: _____

Number sentence: _____

3 $\frac{1}{3} + \frac{1}{12} = ?$

Estimate: _____

Sum: _____

Number sentence: _____

4 $\frac{2}{6} + \frac{1}{4} = ?$

Estimate: _____

Sum: _____

Number sentence: _____

5 $\frac{2}{3} + \frac{1}{4} = ?$

Estimate: _____

Sum: _____

Number sentence: _____

6 $\frac{1}{2} + \frac{1}{5} = ?$

Estimate: _____

Sum: _____

Number sentence: _____

Explaining Place-Value Patterns

① Solve.

   a. 45 * 10 = _____

   b. 45 * 10 * 10 = _____

   c. 45 * 10 * 10 * 10 = _____

   d. 45 * 10 * 10 * 10 * 10 = _____

② Look at your answers to Problem 1.

   a. What pattern do you notice in the number of zeros?

   _____

   _____

   _____

   _____

   b. What pattern do you notice in the value of the products?

   _____

   _____

   _____

   _____

   c. Do you think the patterns will hold true no matter how many 10s are in the problem? Use what you know about place value to explain your answer.

   _____

   _____

   _____

   _____

SRB
68-69,
95-96

3 Solve.

    a. $328 * 10^2 =$ _____

    b. $328 * 10^5 =$ _____

    c. $328 * 10^7 =$ _____

    d. $328 * 10^4 =$ _____

    e. $328 * 10^3 =$ _____

    f. $328 * 10^1 =$ _____

4 Look at your answers to Problem 3.

    a. What pattern do you notice in the number of zeros?

    _____

    _____

    _____

    _____

    b. Use the pattern to help you write a rule for how to multiply a whole number by a power of 10.

    _____

    _____

    _____

    _____

    c. Use what you know about place value to explain why your rule will always work.

    _____

    _____

    _____

    _____

Math Boxes

① Write each number in standard notation.

a. $(3 \times 1,000,000) + (4 \times 100,000) +$
$(2 \times 10,000) + (1 \times 1,000) +$
$(9 \times 100) + (8 \times 10) + (7 \times 1) =$

_____

b. _____ = 8 [10,000s] +
2 [1,000s] + 4 [100s] + 5 [10s] +
6 [1s]

c. $100,000 + 20,000 + 8,000 + 300 +$
$20 + 8 =$ _____

SRB
70

② a. Round 42 to the nearest ten.

_____

b. Round 382 to the nearest hundred.

_____

c. Round 8,461 to the nearest thousand.

_____

d. Round 4.2 to the nearest whole.

_____

SRB
79-82,
126-127

③ Write the money amounts in dollars-and-cents notation.

one dollar and ten cents

_____

three dollars and fifty-two cents

_____

Circle the amount that is greater.

$5.75 $5.57

SRB   SRB
121-123   275

④ Place the numbers 3, 6, 9, 2, and 4 on the number line.

10

0

⑤ Write each fraction as a decimal.

a. $\frac{4}{10}$ _____

b. $\frac{8}{10}$ _____

c. $\frac{52}{100}$ _____

d. $\frac{40}{100}$ _____

SRB
116

⑥ What is the value of the bold digit?

a. $5.\mathbf{4}3 _____

b. $\mathbf{6}.27 _____

c. $8\mathbf{2}.76 _____

d. $9.0\mathbf{2} _____

SRB
118-119

Renaming Fractions and Mixed Numbers

Fractions greater than 1 can be expressed as mixed numbers, such as $2\frac{1}{3}$ and $1\frac{4}{3}$, and as fractions with a numerator larger than the denominator, such as $\frac{7}{3}$. You know several ways to rename fractions as mixed numbers and mixed numbers as fractions.

SRB
171-173

Use fraction circle pieces: Show the original number. Make fair trades between wholes and same-size pieces to rename.

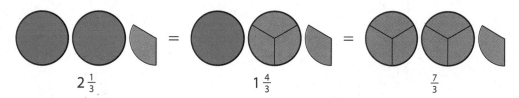

$$2\frac{1}{3} \qquad\qquad 1\frac{4}{3} \qquad\qquad \frac{7}{3}$$

Use number lines: Use fraction names for whole numbers and count up. The number line on the left shows that $2\frac{1}{3}$ is $\frac{4}{3}$ past 1, so $2\frac{1}{3} = 1\frac{4}{3}$. The number line on the right shows that $2 = \frac{6}{3}$, and $2\frac{1}{3}$ and $\frac{7}{3}$ are both $\frac{1}{3}$ past 2, so $2\frac{1}{3} = \frac{7}{3}$.

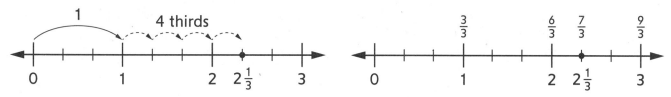

Think about making or breaking wholes:

- To rename $2\frac{1}{3}$ as a fraction, think: *How many thirds are in $2\frac{1}{3}$?* There are 2 wholes. I can break each whole into 3 thirds. Two groups of 3 thirds is the same as 2 * 3 thirds = 6 thirds, or $\frac{6}{3}$. Add one more third to get $\frac{7}{3}$.
- To rename $\frac{7}{3}$ as a mixed number, think: *How many groups of 3 thirds are in 7? What's left over?* There are 2 groups of 3 thirds in 7, with 1 third left over. $\frac{7}{3} = 2\frac{1}{3}$

For Problems 1–3, rename each fraction as a mixed number. Make as many wholes as you can.

1 $\frac{9}{4} =$ _____

2 $\frac{12}{5} =$ _____

3 $\frac{15}{8} =$ _____

For Problems 4–6, write at least two other names with the same denominator for each mixed number.

4 $3\frac{4}{5} =$ _____

5 $2\frac{1}{6} =$ _____

6 $4\frac{1}{2} =$ _____

For Problems 7–9, fill in the missing whole number or missing numerator.

7 $4\frac{2}{5} = \dfrac{\boxed{}}{5}$

8 $\dfrac{\boxed{}}{8} = \dfrac{18}{8}$ (as $\boxed{}\frac{2}{8}$)

9 $2\frac{5}{3} = 3\dfrac{\boxed{}}{3}$

99

Math Boxes

Math Boxes

1 Estimate and solve.

_____
(estimate)

$32\overline{)728}$

$728 \div 32 \rightarrow$ _____

SRB
84,
109-110

2 Use estimation strategies to determine whether the number sentences are true or false.

a. $\frac{3}{4} + \frac{1}{2} < 1$ ◯ True ◯ False

b. $1 - \frac{3}{4} > \frac{1}{2}$ ◯ True ◯ False

c. $\frac{2}{3} + \frac{1}{8} > \frac{1}{2}$ ◯ True ◯ False

d. $\frac{3}{2} + \frac{2}{7} > 1\frac{1}{2}$ ◯ True ◯ False

SRB
181-182

3 Solve. Use fraction circle pieces to help you.

a. $\frac{1}{2} + \frac{1}{4} =$ _____

b. $\frac{1}{2} + \frac{2}{6} =$ _____

c. $\frac{4}{8} + \frac{1}{2} =$ _____

d. $\frac{2}{3} + \frac{1}{6} =$ _____

SRB
166, 189

4 Sophia bought 3 flashlights. Each flashlight cost 5 dollars. The batteries for each flashlight cost 2 dollars. Which expression models this situation?

Fill in the circle next to the best answer.

Ⓐ $(3 * 5) + 2$ Ⓑ $(3 + 2) * 5$

Ⓒ $(5 + 2) / 3$ Ⓓ $(5 + 2) * 3$

SRB
42, 44

5 A cook needs to know the volume of his cupboards. Use the model of the cupboards shown below to estimate the total volume of the cupboards.

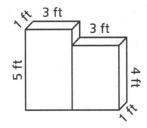

Volume = _____ cubic feet

SRB
233-234

6 There are 24 classes at Lincoln School, each with 23 students.
Write an expression to model how many students go to the school. Then evaluate the expression to find the number of students at the school.

Expression: _____

Answer: _____ students

SRB
38, 44,
100-104

100

Solving Fraction Number Stories

Solve each number story. You can use fraction circle pieces, number lines, pictures, and other tools or models to help you. Show your work.

SRB
30,
178-181

1 Four friends shared 5 sandwiches after their hike. If they each ate an equal share, how many sandwiches did each friend eat?

Answer: _____

_____

2 Josh combined $\frac{1}{2}$ carton of eggs with $\frac{1}{3}$ carton of eggs. He said, "Now I have $\frac{2}{5}$ of a carton." Is Josh correct?

Answer: _____

How do you know?

_____

_____

_____

_____

_____

3 Ryan lives $3\frac{1}{4}$ miles from school. Kayla lives $2\frac{3}{4}$ miles from school. How much farther from school does Ryan live than Kayla?

Answer: _____

_____

4 Delilah was playing *Fraction Capture*. She wrote her initials on a $\frac{1}{3}$ section and a $\frac{1}{6}$ section. What is the sum of the sections she initialed?

Answer: _____

_____

Solving Fraction Number Stories (continued)

SRB
30,
178-181

---

5 Lauren had $\frac{3}{4}$ gallon of paint. She poured in an additional $\frac{1}{8}$ gallon from another can. How much paint does Lauren have?

Answer: _____

_____

---

6 Codyone is cutting fabric for a quilt. She has 4 feet of cloth to make 12 quilt pieces. If she uses all of the fabric, how long should each quilt piece be?

Answer: _____

_____

---

7 Alyssa started with $\frac{7}{8}$ jar of jam. She used about $\frac{1}{16}$ of the jam to make a sandwich. Alyssa said, "I don't need to put jam on the grocery list yet. We still have about $\frac{3}{4}$ jar." Is Alyssa correct?

Answer: _____

How do you know?

_____

_____

_____

_____

---

8 Tyrell and his mom went grocery shopping. They bought $1\frac{1}{6}$ pounds of carrots and $2\frac{5}{6}$ pounds of potatoes. Tyrell carried the carrots and potatoes in one grocery bag. How heavy was the bag?

Answer: _____

_____

Practicing Division

For Problems 1 and 2, make an estimate.
Then divide using partial-quotients division. Report your remainder as a fraction.
Use your estimate to check that your answers make sense.

SRB
108-110,
113-114

① 5,926 / 48 = ?

Estimate: _____

② 9,031 / 71 = ?

Estimate: _____

Answer: _____

Answer: _____

For Problems 3 and 4, write a number model for the story using a letter for the unknown. Then
solve the story. Remember to think about what you should do with the remainder.

③ A food pantry received a donation of
248 cans of chicken. They want to
distribute them evenly to 9 different soup
kitchens in the city. How many cans of
chicken will each soup kitchen get?

Number model: _____

④ Calvin is cutting a roll of paper into 4
pieces to make signs for the school
carnival. The paper is 145 inches long.
How long should he make each banner?

Number model: _____

Answer: _____

Answer: _____

⑤ Explain how you found your answer to Problem 4.

_____

_____

_____

103

Math Boxes

1 Jackson had $\frac{5}{6}$ yard of ribbon. He used $\frac{2}{6}$ yard to decorate a present. How many yards does he have left?

_____
(number model)

Answer: _____ yard

SRB
178-180,
186

2 Solve.

a.   4 7 4
    ×   4

b.   4 7 4
    ×   8

SRB
100,
102, 104

3 Look at each number sentence. Choose True or False.

a. $16 - (3 + 5) = 18$

   ◯ True   ◯ False

b. $(4 + 2) * 5 = 30$

   ◯ True   ◯ False

c. $100 ÷ (25 + 25) + 5 = 7$

   ◯ True   ◯ False

d. $\{(40 - 4) ÷ 6\} + 8 = 14$

   ◯ True   ◯ False

SRB
42-43

4 Use the fractions listed. Fill in the blanks to make the number sentences true.

$\frac{1}{10}$ $1\frac{1}{4}$ $\frac{1}{8}$

a. $\frac{1}{2} +$ _____ < 2

b. $2\frac{1}{3} -$ _____ > 2

c. $\frac{4}{5} +$ _____ < 1

SRB
181-184

5 **Writing/Reasoning** Explain how you know the number sentence you wrote for Problem 4b is true.

_____

_____

_____

_____

SRB
10-11,
181-184

Fraction-Of Problems

Work with a partner or a small group to solve the problems. You can use counters, drawings, or number lines to help you. Be prepared to explain how you solved the problems.

SRB
195

① There are 56 beads on a necklace. $\frac{1}{4}$ of the beads are blue. How many beads are blue?

_____ beads

② Jenna had 45 yards of yarn. She used $\frac{1}{5}$ of it for a knitting project. How much yarn did she use?

_____ yards

③ Morris has a rectangular garden with an area of 60 square feet. $\frac{1}{10}$ of the garden is planted with bean plants. How many square feet are planted with bean plants?

_____ square feet

Try This

④ The length of my living room is 24 feet. The width of my living room is $\frac{1}{2}$ the length. What is the area of my living room?

_____ square feet

Using Models to Estimate Volumes

A mathematical model of each real-world object is given below.

For each object, use the mathematical model to estimate its volume. Be sure to include a unit when you write the volume.

Then write one or more number sentences to show how you found the volume.

 1

10 cm
5 cm 4 cm

Volume: _____

_____
(number sentence)

 2

13 in.
12 in. 12 in.

Volume: _____

_____
(number sentence)

3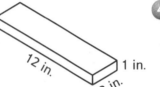

12 in. 1 in.
3 in.

Volume: _____

_____
(number sentence)

4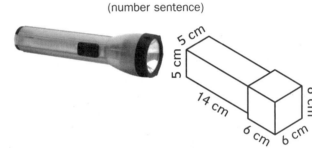

5 cm 5 cm
14 cm
6 cm 6 cm 6 cm

Volume: _____

_____

_____
(number sentences)

5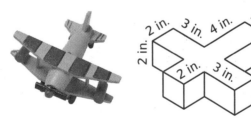

2 in. 2 in. 3 in. 4 in. 2 in. 2 in.
2 in.   2 in.
2 in. 3 in.

Volume: _____

_____

_____
(number sentences)

6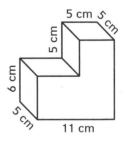

5 cm 5 cm
5 cm
6 cm
5 cm
5 cm 11 cm

Volume: _____

_____

_____
(number sentences)

106

(tl)McGraw-Hill Education/Mark Steinmetz; (tr)Mark Steinmetz; (c)Mark Steinmetz; (cr)McGraw-Hill Education/Mark Steinmetz; (bl)©Jules Frazier/Getty Images; (br)©Comstock Images/Alamy

Math Boxes

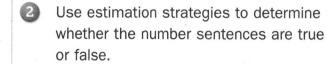

1 Estimate and solve.

_____
(estimate)

$$43\overline{)1{,}298}$$

$1{,}298 \div 43 \rightarrow$ _____

SRB
84,
109-110

2 Use estimation strategies to determine whether the number sentences are true or false.

a. $\frac{2}{3} - \frac{1}{4} < 1\frac{1}{2}$ ○ True   ○ False

b. $\frac{1}{7} + \frac{2}{3} < 1$ ○ True   ○ False

c. $\frac{1}{3} + \frac{3}{4} < \frac{1}{2}$ ○ True   ○ False

d. $\frac{5}{4} - \frac{1}{10} > 1\frac{1}{2}$ ○ True   ○ False

SRB
181-182

3 Solve. Use fraction circle pieces to help you.

a. $\frac{3}{4} + \frac{1}{8} =$ _____

b. $\frac{1}{2} + \frac{5}{8} =$ _____

c. $\frac{2}{5} + \frac{3}{10} =$ _____

d. $\frac{1}{2} + \frac{2}{5} =$ _____

SRB
166, 189

4 Julio bought 4 pounds of carrot sticks and 3 pounds of celery sticks for the school potluck. He had 6 bowls for serving vegetables. He wants to put an equal amount of carrots and celery in each bowl.

Write an expression to show how many pounds of vegetables would be in each bowl.

_____

SRB
42, 44

5 Dixie wants to buy new stuffing for her couch. The drawing below is a model of her couch. Use the model to estimate the volume of her couch.

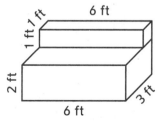

Volume = _____ cubic feet

SRB
233-234

6 Each of the 17 gorillas at the nature preserve eats 45 pounds of food a day. Write an expression that models the amount of food all of the gorillas eat in a day. Then evaluate the expression to find how much food they eat.

Expression: _____

Answer: _____ pounds

SRB
38, 44,
100-104

107

More Fraction-Of Problems

Solve each problem. You can use drawings to help. Be sure to check that your answers make sense.

SRB
195-196

1. What is $\frac{1}{2}$ of 5?

2. What is $\frac{1}{5}$ of 12?

Answer: _____

Answer: _____

3. What is $\frac{1}{4}$ of 2?

4. What is $\frac{1}{8}$ of 6?

Answer: _____

Answer: _____

5. Fredrick went to a farmer's market and bought 8 quarts of strawberries. He wants to keep $\frac{1}{3}$ of the strawberries and give away the rest. How many quarts of strawberries will he keep?

6. Shelby has a 4-pound bag of mixed nuts. She wants to put $\frac{1}{6}$ of the nuts at each of the 6 snack tables at a fundraiser. How many pounds of nuts should she put on each table?

Answer: _____ quarts

Answer: _____ pound

7. Explain how you checked that your answer to Problem 6 made sense.

_____

_____

_____

Math Boxes

1 Emma and her friends hiked $1\frac{1}{3}$ miles on Saturday. The next day they hiked $\frac{2}{3}$ mile. How many miles did they hike in all?

_____
(number model)

Answer: _____ miles

SRB
178-180,
186-187

2 Solve.

a.   3 7 5
    ×   3
   _____

b.   3 7 5
    ×   9
   _____

SRB
100,
102, 104

3 Solve.

a. $(28 / 7) * 3 =$ _____

b. $\{(14 / 7) + (12 / 6)\} * 5 =$ _____

c. $(3 * 10^2) + (5 * 2) =$ _____

d. $32 + \{(8 * 2) / (2 + 2)\} =$ _____

SRB
42-43

4 Use the fractions listed. Fill in the blanks to make each number sentence true.

$\frac{1}{3}$ $\frac{1}{6}$ $\frac{2}{7}$ $2\frac{3}{4}$ $\frac{1}{8}$ $1\frac{2}{3}$

a. _____ + _____ $< \frac{1}{2}$

b. _____ − _____ $> 1\frac{1}{2}$

c. _____ + _____ $> 2\frac{1}{2}$

SRB
181-184

5 **Writing/Reasoning** Write a number story that could be modeled by the number sentence in Problem 3a.

_____

_____

_____

_____

SRB
42-44

Math Boxes

109

Math Boxes

1 Which shows expanded form for the number 942,462?
Choose the best answer.

◯ 94 × 10,000 + 24 × 100 + 62 × 10

◯ 9 [100,000s] + 4 [10,000s] + 2 [1,000s] + 4 [100s] + 6 [10s] + 2 [1s]

◯ 9 [10,000s] + 4 [1,000s] + 2 [100s] + 4 [10s] + 6 [1s] + 2 [0s]

SRB 70

2 a. Round 318 to the nearest ten.

_____

b. Round 4,135 to the nearest hundred.

_____

c. Round 23,891 to the nearest thousand.

_____

SRB 79-82

3 Write the money amounts in dollars-and-cents notation.

ten dollars and fifteen cents

_____

six dollars and eight cents

_____

Circle the amount that is greater.

$217.93 $217.95

SRB 121-123

4 Write the number that each point represents on the number line.

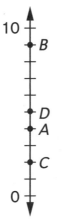

A: _____

B: _____

C: _____

D: _____

SRB 275

5 Write each fraction as a decimal.

a. $\frac{32}{100}$ _____

b. $\frac{9}{10}$ _____

c. $\frac{10}{100}$ _____

SRB 116

6 What is the value of the bold digit?

a. $32.4**2** _____

b. $1**1**6.26 _____

c. $0.8**6** _____

SRB 118-119

110

Math Boxes

① Place the fractions on the number line.

$$\frac{7}{2} \qquad \frac{3}{2} \qquad \frac{5}{2}$$

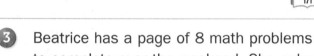

Write $\frac{7}{2}$ as a mixed number. _____

SRB
159-160,
171

② Make an estimate. Then use U.S. traditional multiplication to solve.

_____
(estimate)

$$\begin{array}{r} 3\ 0\ 2 \\ *\quad 8\ 9 \\ \hline \end{array}$$

SRB
83, 103

③ Beatrice has a page of 8 math problems to complete over the weekend. She solved 3 problems on Saturday morning and 2 on Saturday afternoon.

Write a number sentence with fractions that describes how much of the page Beatrice has completed.

_____

What fraction of the page did Beatrice complete on Saturday?

_____

SRB
178-180,
186

④ Solve.

a. $\frac{1}{2}$ of 6 = _____

b. $\frac{1}{2}$ of 8 = _____

c. $\frac{1}{3}$ of 12 = _____

d. $\frac{1}{4}$ of 20 = _____

SRB
195

⑤ **Writing/Reasoning** Choose a fraction from Problem 1. Write a division number story that has that fraction as the answer.

_____

_____

_____

_____

_____

SRB
163-164

Reading and Writing Decimals

SRB
117-118

Ones 1s 1s	.	Tenths 0.1s $\frac{1}{10}$s	Hundredths 0.01s $\frac{1}{100}$s	Thousandths 0.001s $\frac{1}{1,000}$s
	.			
	.			
	.			
	.			
	.			
	.			
	.			
	.			
	.			
	.			
	.			

Write the following decimals in words. Use the place-value chart on journal
page 112 to help you.

SRB
116-119

1 0.67 _____

2 3.8 _____

3 3.622 _____

4 0.804 _____

Write each decimal using numerals. Record them on the place-value chart on page 112.
Then write the value of 4 in each decimal.

5 **a.** four and eight tenths _____   **b.** 4 is worth _____.

6 **a.** forty-eight hundredths _____   **b.** 4 is worth _____.

7 **a.** forty-eight thousandths _____   **b.** 4 is worth _____.

8 **a.** six and four hundred eight thousandths _____

   b. 4 is worth _____.

Rewrite each decimal as a fraction.

9 0.6 _____   10 0.03 _____   11 0.008 _____

Rewrite each fraction as a decimal.

12 $\frac{2}{10}$ _____   13 $\frac{65}{1,000}$ _____   14 $\frac{402}{1,000}$ _____

15 Use the clues to write the mystery
    number.

    Write 5 in the tenths place.

    Write 6 in the ones place.

    Write 2 in the thousandths place.

    Write 1 in the hundredths place.

    _____.___ ___ ___

16 Make the following changes to the
    number 7.849:

    Make the 7 worth $\frac{1}{10}$ as much.

    Make the 8 worth 10 times as much.

    Make the 4 worth $\frac{1}{10}$ as much.

    Make the 9 worth 10 times as much.

    _____.___ ___ ___

Representing Decimals

Math Message

SRB
116-118,
120

1 Write 0.43 in words. _____

2 Write 0.43 in the place-value chart below.

Ones	.	Tenths	Hundredths	Thousandths
	.			

3 The grid below represents 1. Shade the grid to show 0.43.

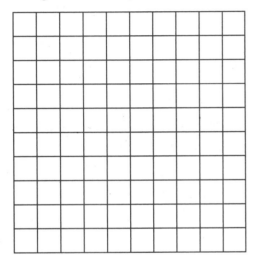

For Problems 4–6, use words, fractions, equivalent decimals, or other representations to write at least three names for each decimal in the name-collection box. Then shade the grid to show the decimal.

4

0.8

114

Representing Decimals (continued)

5

0.620

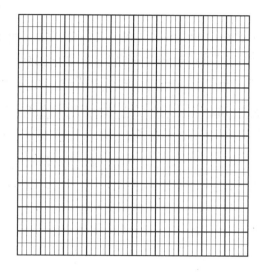

6

0.418

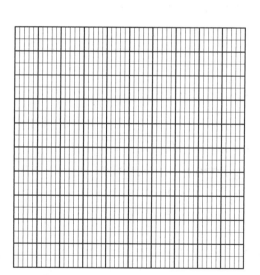

Fraction Number Stories

For each story, write a number model with a letter for the unknown. Then solve the story. You can draw pictures or use fraction circle pieces to help.

SRB
163-164,
178-180

1 A chef divides 5 heads of lettuce equally among 12 salad plates. How much lettuce will be on each plate?

Number model: _____

Answer: _____ head of lettuce

2 Alma has a 10-foot roll of wrapping paper. She cuts off $2\frac{3}{4}$ feet of paper to wrap a gift. How much paper is left on the roll?

Number model: _____

Answer: _____ feet of paper

3 Vernon mixed $2\frac{1}{3}$ cups of water with $2\frac{1}{3}$ cups of white vinegar to make a cleaning solution. How much cleaning solution did he make?

Number model: _____

Answer: _____ cups

4 In a relay race one runner ran $4\frac{3}{10}$ miles. The next runner ran $4\frac{9}{10}$ miles. What is the total distance they ran?

Number model: _____

Answer: _____ miles

5 A guinea pig weighs $1\frac{7}{8}$ pounds. A rabbit weighs $3\frac{3}{8}$ pounds. How much more does the rabbit weigh than the guinea pig?

Number model: _____

Answer: _____ pounds

6 A bicycle relay race is 24 miles long. Nikita's team has 7 members who will each ride the same distance in the race. How far will each team member ride?

Number model: _____

Answer: _____ miles

Math Boxes

1 Write each number in expanded form.

a. 21,756,834

_____

_____

b. 311,019

_____

_____

SRB
70

2 Estimate. Then use partial-quotients division to solve.

$2{,}731 \div 31 = ?$

_____
(estimate)

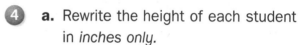

$2{,}731 \div 31 \rightarrow$ _____

SRB
84,
109-110

3 Find the volume of the rectangular prism. Use the formula $V = l * w * h$.

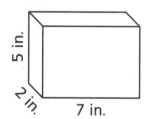

5 in.

2 in.   7 in.

$V =$ _____
(number model)

$V =$ _____ in.³

SRB
233

4 **a.** Rewrite the height of each student in *inches only.*

Juan: 5 feet, 4 inches = _____

Marilu: 4 feet, 11 inches = _____

b. How many inches taller than Marilu is Juan?

_____ inches

SRB
215-217,
219, 328

5 Write >, <, or =.

a. $\frac{2}{3} - \frac{1}{2}$ _____ $\frac{1}{2}$

b. $\frac{3}{4} + \frac{3}{8}$ _____ 1

c. $\frac{3}{4} - \frac{1}{2}$ _____ $\frac{1}{4}$

SRB
181-182

6 The swimming pool is open 8 hours a day. The pool manager must divide pool time equally among 5 groups: camp, swim team, swim lessons, family swim, and open swim. How much time should she allow for each group?

_____ hours

SRB
163-164

Math Boxes

117

Writing Decimals in Expanded Form

Math Message

Shade the grid to represent 0.3. Add shading to the grid in another color so that the grid shows 0.31 in all. Then use a third color to add shading so that the grid shows 0.312.

SRB
118, 120

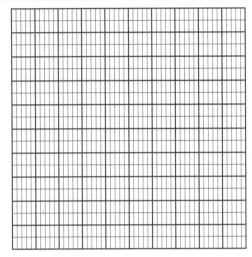

Use different versions of expanded form to complete the table below.

Standard Notation	Versions of Expanded Form		
	Sum of Decimals in Standard Notation	Sum of Multiplication Expressions (Decimals)	Sum of Multiplication Expressions (Fractions)
Example: 0.568	$0.5 + 0.06 + 0.008$	$(5 * 0.1) + (6 * 0.01) + (8 * 0.001)$	$5 * \frac{1}{10} + 6 * \frac{1}{100} + 8 * \frac{1}{1,000}$
2.473			$(2 * 1) + \left(4 * \frac{1}{10}\right) + \left(7 * \frac{1}{100}\right) + \left(3 * \frac{1}{1,000}\right)$
0.094			
7.752			
	$0.6 + 0.03 + 0.007$		

Representing Decimals in Expanded Form

Use three number cards to create a decimal on your decimal place-value mat. Record the decimal you created in one of the boxes below. Write the decimal in expanded form. Then shade a thousandths grid using a different color to show the value of each digit. Repeat to complete all four boxes.

Decimal: 0. _____ _____ _____

Expanded form: _____

_____

Decimal: 0. _____ _____ _____

Expanded form: _____

_____

Decimal: 0. _____ _____ _____

Expanded form: _____

_____

Decimal: 0. _____ _____ _____

Expanded form: _____

_____

Math Boxes

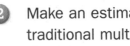

1 Label each mark on the number line with the appropriate fraction below.

$$\frac{4}{3} \qquad \frac{3}{2} \qquad \frac{7}{4}$$

1   2

Write $\frac{4}{3}$ as a mixed number. _____

SRB
159-160,
171

2 Make an estimate. Then use U.S. traditional multiplication to fill in the missing numbers.

_____
(estimate)

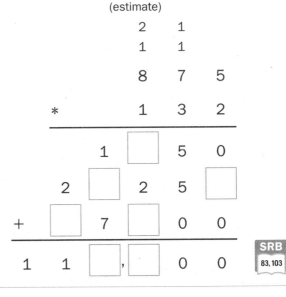

SRB
83, 103

3 Dawn wants to plant vegetables and herbs in her garden. If she plants herbs in $\frac{3}{8}$ of the garden, how much of the garden is left over for vegetables?

(number model)

Answer: _____ garden

SRB
178-180,
186

4 What is:

a. $\frac{1}{3}$ of 24? _____

b. $\frac{1}{4}$ of 24? _____

c. $\frac{1}{6}$ of 24? _____

SRB
195

5 **Writing/Reasoning** Explain how you solved Problem 4c.

_____

_____

_____

_____

SRB
195

120

Interpreting Remainders

Solve each number story. Show your work. Explain what you decided to do with the remainder.

SRB
109,
113-114

1   Bre earned 189 tickets playing different games at the fair. If each prize costs 15 tickets, how many prizes can Bre get?

Number model: _____

Quotient: _____ Remainder: _____

Answer: Bre can get _____ prizes.

Circle what you did with the remainder.

    Ignored it   Reported it   Rounded the
                       as a fraction   quotient up

Why? _____

_____

2   Elisbeth is 58 inches tall. What is her height in feet?

Remember: 1 foot = 12 inches

Number model: _____

Quotient: _____ Remainder: _____

Answer: Elisbeth is _____ feet tall.

Circle what you did with the remainder.

    Ignored it   Reported it   Rounded the
                       as a fraction   quotient up

Why? _____

_____

Math Boxes

1 Write the following number in expanded form.

3,768,412,000

_____

_____

_____

_____

_____

SRB
70

2 Estimate. Then use partial-quotients division to solve.

$8,096 \div 21 = ?$

_____
(estimate)

$8,096 \div 21 \rightarrow$ _____

SRB
84,
109-110

3 Find the volume of the rectangular prism. Use the formula $V = B * h$.

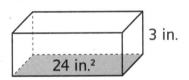

3 in.

24 in.²

$V =$ _____ cubic inches

SRB
233

4 **a.** Rewrite each baby's birth weight in ounces (oz) only.

Remember: 16 oz = 1 lb

Sadie: 5 lb, 11 oz = _____ oz

Kevin: 8 lb, 9 oz = _____ oz

b. How many more ounces did Kevin weigh at birth than Sadie?

_____ oz

SRB
215-217

5 Write >, <, or =.

a. $1\frac{1}{3} - \frac{2}{3}$ _____ 1

b. $\frac{3}{4} - \frac{1}{8}$ _____ $\frac{1}{2}$

c. $\frac{4}{8} + \frac{6}{12}$ _____ 1

SRB
181-182

6 Matt has a job as a dog walker. He has 4 hours to walk 18 dogs. Which expressions represent the amount of time that should be spent walking each dog? Select all that apply.

☐ $\frac{18}{4}$ hours   ☐ $(18 \div 4)$ hours

☐ $\frac{4}{18}$ hour   ☐ $(4 \div 18)$ hour

SRB
163-164

Math Boxes

Identifying the Closer Number

Math Message

SRB
124-125

1. Shade the grid at the right to show 0.28.

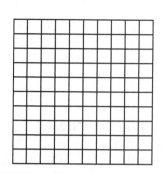

2. Is 0.28 closer to 0.2 or 0.3? Use the grid to help you decide. Be ready to explain your reasoning.

   0.28 is closer to _____.

---

3. Label the number line below to show whether 0.28 is closer to 0.2 or 0.3.

   0.2   0.3

4. Shade the grid at the right to show 0.619.
   Use the grid to help you solve Problem 5.

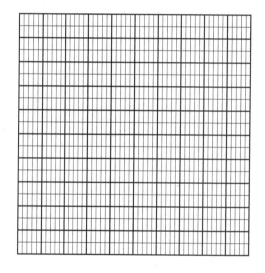

5. **a.** Between which two hundredths is 3.619?

   3.619 is between _____ and _____.

   b. What number is exactly halfway between the numbers you wrote in Problem 5a?

   _____

6. Round 3.619 to the nearest hundredth. Use your answers from Problems 4 and 5 and the number line below to help you.

   _____

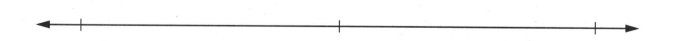

Rounding Decimals

Use the number lines to round each number. Be sure to label the tick marks on each number line.

1 Round 3.6 to the nearest whole number.

SRB
125

Rounded number: _____ Did you round up or down? _____

2 Round 2.73 to the nearest tenth.

Rounded number: _____ Did you round up or down? _____

3 Round 2.73 to the nearest whole number.

Rounded number: _____ Did you round up or down? _____

4 Round 4.254 to the nearest hundredth.

Rounded number: _____ Did you round up or down? _____

5 Round 4.254 to the nearest tenth.

Rounded number: _____ Did you round up or down? _____

6 Round 4.254 to the nearest whole number.

Rounded number: _____ Did you round up or down? _____

Rounding Decimals in Real-World Contexts

Read about different real-world situations below and round the decimals as directed. Use a number line or grids to help you, if needed.

SRB
124-127

1. At the district track meet each running event is timed to the nearest thousandth of a second using an electronic timer. However, the district's track rules require times to be reported with only 2 decimal places. Round each time to the nearest hundredth of a second.

Note: sec = second(s)

Electronic Timer	Reported Time	Electronic Timer	Reported Time
a. 10.752 sec	sec	**b.** 55.738 sec	sec
c. 16.815 sec	sec	**d.** 43.505 sec	sec
e. 20.098 sec	sec	**f.** 52.996 sec	sec

Explain how you rounded 20.098 to the nearest hundredth.

_____

_____

_____

2. Supermarkets often show unit prices for items. This helps customers compare prices to find the best deal. A unit price is found by dividing the price of an item (in cents or dollars and cents) by the quantity of the item (often in ounces or pounds). When the quotient has more decimal places than are needed, some stores round to the nearest tenth of a cent.

Example: A 16 oz container of yogurt costs $3.81.

- $3.81 * 100 cents per dollar = 381¢
- 381¢ ÷ 16 oz = 23.8125¢ per ounce
- 23.8125¢ is rounded down to 23.8¢ per ounce

Round each unit price to the nearest tenth of a cent.

a. 28.374¢ _____¢   **b.** 19.756¢ _____¢

c. 16.916¢ _____¢   **d.** 20.641¢ _____¢

e. 18.459¢ _____¢   **f.** 21.966¢ _____¢

Math Boxes

Math Boxes

1 Keegan practices karate for 60 minutes every weekday. If the summer has 47 weekdays, how many minutes will he have practiced by the end of summer?

_____
(number model)

_____
(estimate)

Answer: _____

SRB
44, 83,
100-104

2 Shade the grid to represent the decimal 0.8.

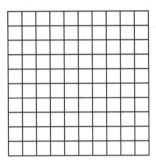

Write the decimal in words.

_____

SRB
117, 120

3 Circle the benchmark that is closest to each sum or difference.

a. $\frac{3}{8} + \frac{9}{10}$

  0      $\frac{1}{2}$ 1 $1\frac{1}{2}$ 2

b. $1\frac{1}{6} - \frac{3}{5}$

  0      $\frac{1}{2}$ 1 $1\frac{1}{2}$ 2

SRB
181-184

4 Olivia is buying bundles of wood for a campfire. The bundles of wood can't be split up. If each bundle costs $3, how many bundles can she buy with $10?

_____
(number model)

Answer: _____ bundles of wood

SRB
44, 113

5 **Writing/Reasoning** Explain how you solved Problem 3b.

_____

_____

_____

_____

SRB
181-184

126

Locating Cities on a Map of Ireland

Bantry	B-1	Dublin	F-4	Lahinch	B-4	Omagh	E-7
Belfast	F-7	Dundalk	F-6	Larne	F-7	Tralee	B-2
Carlow	E-3	Galway	C-4	Limerick	C-3	Tuam	C-5
Castlebar	B-6	Gort	C-4	Mullingar	E-5	Westport	_____
Derry	E-8	Kilkee	B-3	Navan	E-5	Wicklow	_____

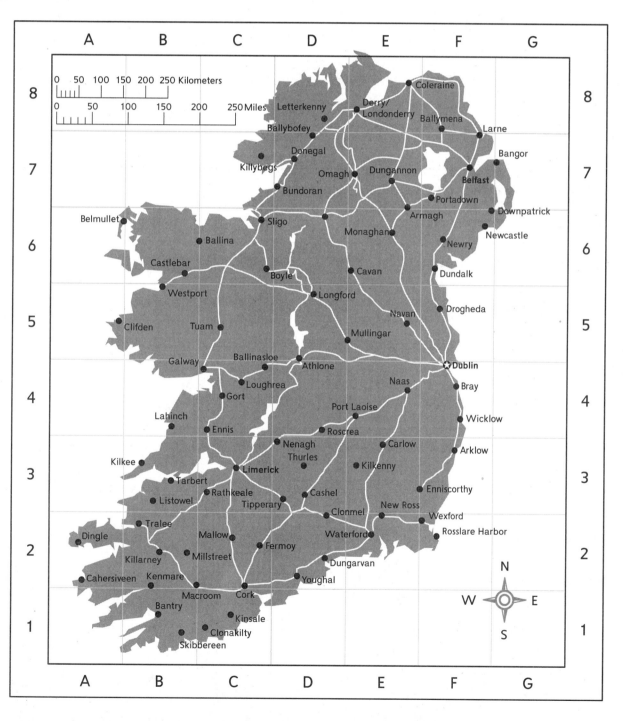

Plotting Points on a Coordinate Grid

SRB
275

1 Use the grid shown below as you follow directions from your teacher.

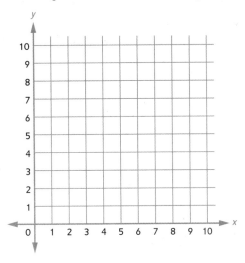

For Problems 2–6, use the coordinate grid shown below.

2 Write the ordered pairs for:

   a. Point A _____   **b.** Point B _____

   c. Point C _____   **d.** the origin _____

3 Use a straightedge. Connect points A, B, and C in order.

4 On the same grid, plot and label the points listed below.

   D (6, 10)   E (9, 10)

   F (10, 9)   G (10, 6)

   H (9, 5)   J (6, 5)

   K (5, 6)

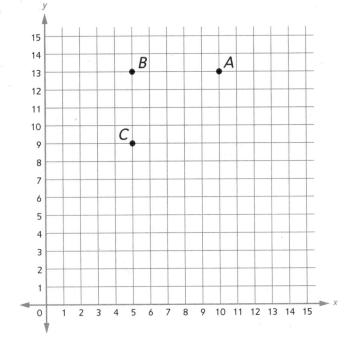

5 Use a straightedge. Connect the points in alphabetical order beginning with point C.

6 What image did you create by connecting the points?

   _____

How Much Soil?

1 **a.** David helped his father build a large planter to grow flowers in front of their house. How much soil can the planter hold?

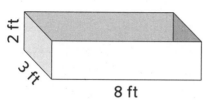

The planter can hold _____ cubic feet of soil.

Number model: _____

b. David found 11 unopened bags of potting soil in the garage. Each bag contained 2 cubic feet of soil. David dumped all the bags of soil into the planter. Draw a line on the picture above to show about how much soil is in the planter. Explain how you figured it out.

_____

_____

_____

2 David's mother used scraps of wood to build the vegetable planter shown below for the back patio.

a. How much soil can the planter hold?

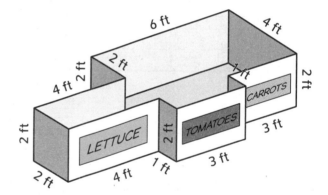

The vegetable planter can hold _____ cubic feet of soil.

b. Explain how you solved Part a.

_____

_____

_____

_____

c. If the planter doesn't fit on the patio, David's mom will remove the lettuce section.

Then how much soil would be needed to fill the planter? _____ cubic feet

3 Talk to a partner about how you solved each problem. Compare your strategies.

Math Boxes

Math Boxes

1 Sasha walked $\frac{3}{5}$ mile to school. After school, she walked another $\frac{1}{5}$ mile to ballet practice. From ballet, she walked $\frac{4}{5}$ mile home. How far did Sasha walk all together?

_____
(number model)

Answer: _____

SRB
178-180,
186

2 Fill in the name-collection box with at least 3 names.

0.25

SRB
116-118

3 Write 4.628 in expanded form.

_____

_____

SRB
118

4 Circle the situation below that would have the answer: $\frac{8}{5}$ kilograms of krill.

A. 8 penguins eat 5 kilograms of krill. How much does each penguin eat?

B. 5 penguins eat 8 kilograms of krill. How much does each penguin eat?

SRB
163-164

5

This figure models the steps to Shelby's porch. Which of the following are true? Fill in the circle next to <u>all</u> that apply.

○ **A.** The total volume of the steps is 84 ft³.

○ **B.** The volume of step A is 28 ft³.

SRB
233-234

○ **C.** The volume of step B is 60 ft³.

6 Make an estimate. Then solve.

$4{,}211 \div 68$

_____
(estimate)

$4{,}211 \div 68 \rightarrow$ _____

SRB
84,
109-110

Town Map

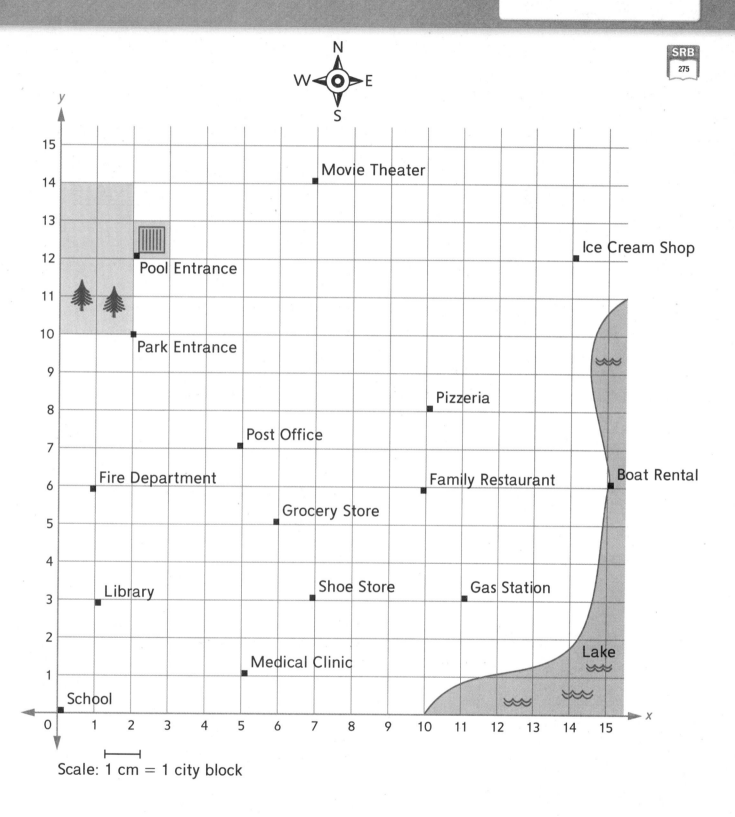

Scale: 1 cm = 1 city block

131

SRB
83,
102-103

For Problems 1–3, make an estimate. Then solve using U.S. traditional multiplication.

1 Estimate:

_____

```
        2 5
    *   1 1
    _____
```

2 Estimate:

_____

```
        4 3 2
    *     2 9
    _____
```

3 Estimate:

_____

```
        2 1 7
    *   3 0 9
    _____
```

Another student estimated and began solving Problems 4–6 using U.S. traditional multiplication. Finish solving the problems.

4 Estimate:

$40 * 60 = 2,400$

```
        1
        3 7
    *   6 2
    _____
        7 4
```

5 Estimate:

$500 * 100 = 50,000$

```
      7   1
      4 9 2
    *   9 8
    _____
      3 6
```

6 Estimate:

$500 * 200 = 100,000$

```
        5 1 1
    *   2 1 9
    _____
              9
```

7 Stephen solved the problem below using U.S. traditional multiplication. He can tell from his estimate that his answer is wrong. Find Stephen's mistake and explain how he could fix it.

Estimate:

$700 * 80 = 56,000$

```
          5   1
          5   1
          6 7 2
    *       8 7
    _____
      4 7 0 4
  + 5 3 7 6
  _____
  1 0,0 8 0
```

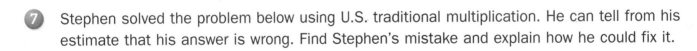

_____

_____

_____

_____

_____

_____

Math Boxes

1 Colette is filling 400 water balloons for a school picnic. Balloons come in bags of 25, 50, 150, and 250. Which set of bags will provide at least 400 balloons, with the fewest number of balloons left over?

Fill in the circle next to the best answer.

○ **A.** 15 bags of 25 balloons

○ **B.** 8 bags of 50 balloons

○ **C.** 3 bags of 150 balloons

○ **D.** 2 bags of 250 balloons

SRB
97-104

2 Shade in the grid to represent the decimal 0.08.

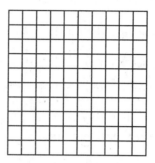

Write the decimal in words.

_____

SRB
117, 120

3 Circle the benchmark that is closest to each sum or difference.

a. $\frac{5}{9} - \frac{3}{6}$

　0　　　$\frac{1}{2}$　　　1

b. $\frac{1}{2} + \frac{5}{6}$

　0　　　$\frac{1}{2}$　　　1

SRB
181-184

4 Lola has 17 photos to place in the school newsletter. She can fit 4 photos per page. How many pages long must the newsletter be to fit all the photos?

_____
(number model)

Quotient: _____ Remainder: _____

Answer: _____ pages

SRB
44, 113

5 **Writing/Reasoning** Explain how you decided what to do with the remainder for Problem 4.

_____

_____

_____

_____

SRB
113

Math Boxes

Graphing Sailboats

1. Find the column labeled Original Sailboat in the table below. Plot the ordered pairs listed in the column on the grid titled Original Sailboat on the next page. Connect the points in the same order that you plot them. You should see the outline of a sailboat.

    SRB 275

2. **a.** Fill in the missing coordinates for New Sailboat 1.

    b. How do you think New Sailboat 1 will be different from the Original Sailboat? Record a conjecture at the top of the column.

    c. Plot the ordered pairs for New Sailboat 1 on the next page. Connect the points in the same order that you plot them.

Original Sailboat	New Sailboat 1 — Rule: Double each number of the original pair.	New Sailboat 2 — Rule: Double the first number of the original pair. Leave the second number the same.	New Sailboat 3 — Rule: Double the second number of the original pair. Leave the first number the same.
Conjecture:			
(8, 1)	(16, 2)	(16, 1)	(8, 2)
(5, 1)	(10, 2)	(10, 1)	(5, 2)
(5, 7)	(10, 14)	(10, 7)	(5, 14)
(1, 2)	(,)	(,)	(,)
(5, 1)	(,)	(,)	(,)
(0, 1)	(,)	(,)	(,)
(2, 0)	(,)	(,)	(,)
(7, 0)	(,)	(,)	(,)
(8, 1)	(,)	(,)	(,)

    d. Complete steps 2a–2c for New Sailboat 2.

    e. Complete steps 2a–2c for New Sailboat 3.

    Be sure to apply each rule to the coordinates from the **Original Sailboat.**

SRB
10-11,
275

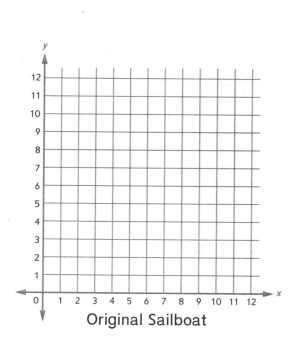

Original Sailboat

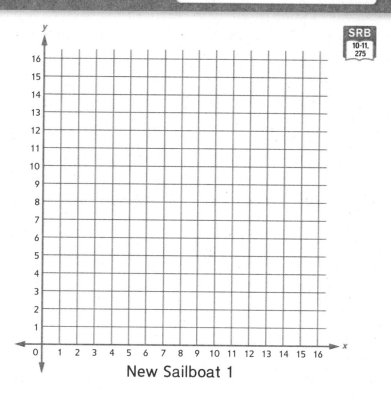

New Sailboat 1

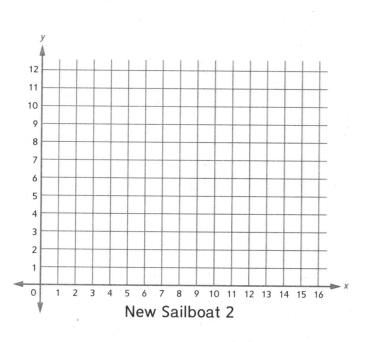

New Sailboat 2

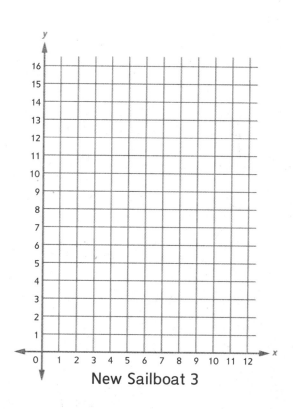

New Sailboat 3

A New Sailboat Rule

1. Circle the rule for Sailboat 4 given to you by your teacher.

   • Triple the first number of the ordered pair.

   • Triple the second number of the ordered pair.

   • Double the first number of the ordered pair; halve the second number.

   • Halve the first number of the ordered pair; double the second number.

   • Other: _____

   _____

2. Make a conjecture about what New Sailboat 4 will look like.

   _____

   _____

3. Create ordered pairs for New Sailboat 4 based on the rule. Write them in the table below.

4. Plot the new set of ordered pairs and connect the points in the order they were plotted.

5. Was your conjecture correct? Explain. _____

   _____

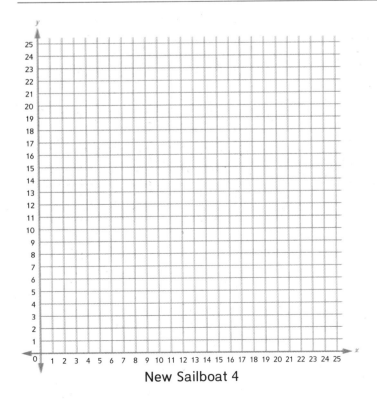

New Sailboat 4

Original Sailboat	New Sailboat 4
(8, 1)	(,)
(5, 1)	(,)
(5, 7)	(,)
(1, 2)	(,)
(5, 1)	(,)
(0, 1)	(,)
(2, 0)	(,)
(7, 0)	(,)
(8, 1)	(,)

Math Boxes

① Dakota had a kite with $25\frac{6}{12}$ feet of string. The kite got stuck in a tree. She lost $4\frac{3}{12}$ feet of string trying to free it. How much string does Dakota still have?

_____
(number model)

Answer: _____

SRB
178-180

② Complete the name-collection box.

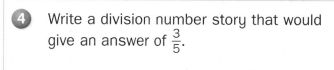

3.09

SRB
116-118

③ Write the number 17.803 in expanded form.

_____

_____

SRB
118

④ Write a division number story that would give an answer of $\frac{3}{5}$.

_____

_____

_____

_____

SRB
163-164

⑤ A clerk stacked these boxes to create a store display. Each box is 1 cubic unit. Find the volume of the store display.

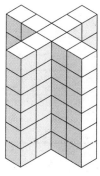

SRB
231-232,
234

$V =$ _____ units^3

⑥ Write a division problem for which your estimate might be:

$7,000 \div 100 = 70$

_____ ÷ _____ → _____

Solve your problem:

SRB
84,
109-110

137

Graphing Data as Ordered Pairs

The data in the table show Lilith's and Noah's ages at 5 different times in their lives.

SRB
55-56,
275

Lilith's Age (years)	Noah's Age (years)
5	1
7	3
9	5
11	7
12	8

Ordered Pairs:

(_____, _____)

(_____, _____)

(_____, _____)

(_____, _____)

(_____, _____)

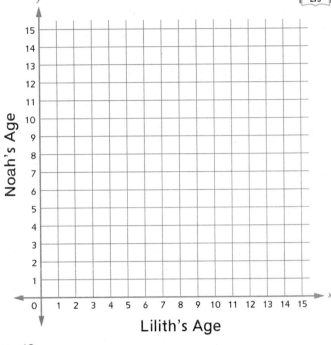

1 Write their ages as ordered pairs. Then plot the points on the grid.

2 What do you notice about the points you plotted?

_____

3 Connect the points with a line. What other information can we get from this line?

_____

4 Use the line to determine the ages of Lilith and Noah at various points in their lives.

    a. When Lilith was 8 years old, Noah was _____ years old.

    b. When Noah was 6 years old, Lilith was _____ years old.

    c. How old will Lilith be when Noah is 11? _____

5 Explain how you solved Problem 4c.

_____

_____

6 Who is older, Lilith or Noah? _____ How much older? _____

Forming and Graphing Ordered Pairs

For each data set, fill in the missing values and write the data as ordered pairs. Plot the points on the grid and connect them using a straightedge. Use the graph to answer the questions.

SRB
55-56, 275

1. Dean is raising money for charity. He earns $2 for each lap he runs around the gym.

Laps Run (x)	$ Earned (y)
1	2
2	
	6
4	

Ordered pairs:

(____, ____)

(____, ____)

(____, ____)

(____, ____)

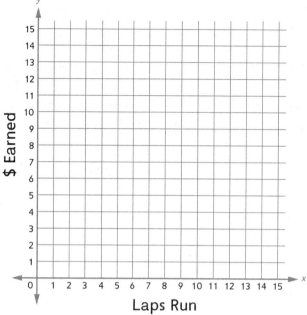

a. If Dean has earned $14, how many laps has he run? _____ laps

b. Put an X on the point on the grid that shows your answer to Part a.

c. What are the coordinates for this point? (_____, _____)

2. Sally uses 2 paintbrushes for each paint jar.

Brushes (x)	Paint Jars (y)
2	1
4	
	3
8	

Ordered pairs:

(____, ____)

(____, ____)

(____, ____)

(____, ____)

a. If Sally uses 6 jars of paint, how many brushes does she need? _____ brushes

b. Put an X on the point on the grid that shows your answer to Part a.

c. What are the coordinates for the point you marked with an X? (_____, _____)

139

Math Boxes

Math Boxes

1 Write < or > to make true number sentences.

    a. 0.5 _____ 1.0

    b. 3.2 _____ 3.02

    c. 4.83 _____ 4.8

    d. 6.25 _____ 6.4

    e. 0.7 _____ 0.07

SRB
121-123

2 Write in standard notation.

$2 \times 10^{3} =$ _____

$7 \times 10^{5} =$ _____

$3 \times 10^{2} =$ _____

SRB
68 69

3 Solve.

$\frac{1}{2}$ of 9 = _____

$\frac{1}{4}$ of 5 = _____

SRB
195

4 Write each decimal in words.

    a. 0.16

    _____

    _____

    b. 3.28

    _____

    _____

SRB
117

5 **Writing/Reasoning** Explain how you compared the decimals in Problem 1.

_____

_____

_____

_____

_____

SRB
121-123

140

Logo

Amy is designing a logo for her school club. She plans to put a trapezoid around the letters RC, which stand for Running Club. Below is the picture of the original trapezoid she drew on a coordinate grid.

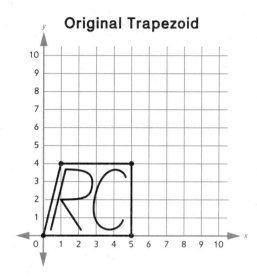

Original Trapezoid

Amy decides she wants to include the school's name, so she needs to make the trapezoid wider. She does not want it to be taller. She developed a rule to help her fix the drawing.

Amy's Rule: Double the first coordinate of all the points.

1. If Amy uses her rule, what do you think the new trapezoid will look like? Why? Be specific in your description.

141

SRB
55-56,
275

② Use Amy's rule to write the coordinates for the new trapezoid.

Original Trapezoid	New Trapezoid
(0, 0)	
(1, 4)	
(5, 4)	
(5, 0)	
(0, 0)	

③ Plot the new coordinates on the grid below. Connect the points in the same order you plot them.

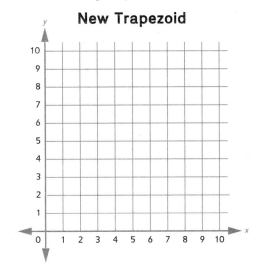

New Trapezoid

④ Does the new trapezoid look the way you expected? Why or why not? Be specific about how it changed.

Math Boxes: Preview for Unit 5

Math Boxes

---

1 Use the unit squares to find the area of the rectangle.

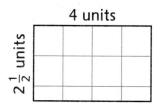

4 units

$2\frac{1}{2}$ units

Area = _____ units²

SRB
224-225

---

2 Write 5 multiples of 7.

_____

_____

SRB
72

---

3 Shoshana and Ariel got different answers on a fraction estimation problem. For the problem $\frac{8}{9} + \frac{1}{4}$:

Shoshana wrote $\frac{8}{9} + \frac{1}{4} > 1$.

Ariel wrote $\frac{8}{9} + \frac{1}{4} < 1$.

Who is correct? _____

How do you know?

_____

_____

_____

SRB
181-182

---

4 Write all the factors of 30.

_____

_____

SRB
73

---

5 Fill in the missing number.

a. $\frac{1}{2} = \frac{5}{\boxed{}}$

b. $\frac{2}{3} = \frac{\boxed{}}{12}$

c. $\frac{9}{\boxed{}} = \frac{90}{100}$

SRB
166,
168-170

---

6 Solve.

a. If 4 is $\frac{1}{2}$ of the whole, what is the whole? _____

b. If 2 is $\frac{1}{3}$ of the whole, what is the whole? _____

SRB
195

143

Decimal Addition and Subtraction with Grids

For Problems 1 and 2:

SRB
129

- Shade the grid in one color to show the first addend.

- Shade more of the grid in a second color to show the second addend.

- Write the sum to complete the number sentence.

 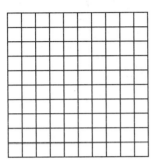

 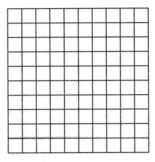

0.6 + 0.22 = _____

0.18 + 0.35 = _____

For Problems 3 and 4:

- Shade the grid to show the starting number.

- Cross out or shade darker to show what is being taken away.

- Write the difference to complete the number sentence.

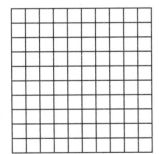

 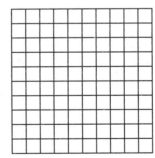

0.47 − 0.20 = _____

0.74 − 0.36 = _____

5 Choose one of the problems above. Clearly explain how you solved it.

_____

_____

_____

_____

Math Boxes

1 Plot the following points on the grid.

a. (1, 1) b. (2, 3)

c. (5, 3) d. (4, 1)

e. (1, 1)

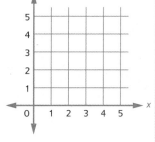

Connect the points in the order given.
What shape have you drawn?

_____

SRB
268, 275

2 Solve. You may use fraction circle pieces to help you.

a. $\frac{1}{2} + \frac{1}{4} = $ _____

b. $\frac{1}{2} + \frac{3}{4} = $ _____

c. $\frac{3}{4} + \frac{1}{8} = $ _____

d. $\frac{1}{4} + \frac{3}{8} = $ _____

SRB
166, 189

3 Make an estimate and then solve.

_____
(estimate)

```
    1  9  4
 *  2  1  5
 _____
```

SRB
83,
100-104

4 Round to the nearest tenth.

a. 45.52 = _____

b. 60.18 = _____

c. 123.45 = _____

d. 38.27 = _____

e. 56.199 = _____

SRB
124-127

5 **Writing/Reasoning** Explain why the order of the numbers in an ordered pair is important.

_____

_____

_____

_____

SRB
275

145

Using Algorithms to Add Decimals

For Problems 1–6, make an estimate. Write a number sentence to show how you estimated. Then solve using partial-sums addition, column addition, or U.S. traditional addition. Show your work. Use your estimates to check that your answers make sense.

SRB
128, 130

1 2.3 + 7.6 = ?	**2** 6.4 + 8.7 = ?	**3** 7.06 + 14.93 = ?
_____ (estimate)	_____ (estimate)	_____ (estimate)
2.3 + 7.6 = _____	6.4 + 8.7 = _____	7.06 + 14.93 = _____
4 21.47 + 9.68 = ?	**5** 3.514 + 5.282 = ?	**6** 19.046 + 71.24 = ?
_____ (estimate)	_____ (estimate)	_____ (estimate)
21.47 + 9.68 = _____	3.514 + 5.282 = _____	19.046 + 71.24 = _____

7 Choose one problem. Answer the questions below.

   a. How did you make your estimate?

   _____

   _____

   b. How did you use your estimate to check that your answer made sense?

   _____

   _____

   _____

146

Math Boxes

Math Boxes

---

1 Put the following numbers in order from least to greatest.

7.1   7.01   0.0071   0.71

_____, _____, _____, _____
Least   Greatest

SRB
121-123

---

2 Write in exponential notation.

a. 30,000 = _____ × $10^{\square}$

b. 6,000,000 = _____ × $10^{\square}$

SRB
68-69

---

3 Evelyn is going on a hike. She will hike 4 miles in all. So far, she has hiked $\frac{1}{2}$ of the total distance. How far has she hiked?

Answer: _____ miles

SRB
195

---

4 Write the decimal in words.

a. 31.04

_____

_____

b. 6.208

_____

_____

SRB
117

---

5 **Writing/Reasoning** Sarah said that 10^4 is the same as 40, because 10 * 4 is 40. Explain Sarah's mistake.

_____

_____

_____

_____

SRB
68

147

Using Algorithms to Subtract Decimals

For Problems 1–6, make an estimate. Write a number sentence to show how you estimated. Then solve using trade-first subtraction, counting-up subtraction, or U.S. traditional subtraction. Show your work. Use your estimates to check that your answers make sense.

SRB
128,
131-132

① 4.6 − 3.2 = ?	② 13.1 − 8.7 = ?	③ 6.87 − 2.52 = ?
_____ (estimate)	_____ (estimate)	_____ (estimate)
4.6 − 3.2 = _____	13.1 − 8.7 = _____	6.87 − 2.52 = _____
④ 24.07 − 12.68 = ?	⑤ 62.432 − 19.712 = ?	⑥ 17.41 − 6.274 = ?
_____ (estimate)	_____ (estimate)	_____ (estimate)
24.07 − 12.68 = _____	62.432 − 19.712 = _____	17.41 − 6.274 = _____

⑦ Choose one problem. Think about the algorithm you used. Answer the questions below.

   a. How did your choice of algorithm help you get an accurate answer?

_____

_____

_____

   b. Was your choice of algorithm the most efficient choice? Why or why not?

_____

_____

_____

Math Boxes

① Write the coordinates for each of the points on the coordinate grid.

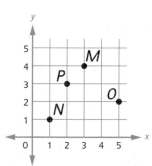

M: (____, ____)

N: (____, ____)

O: (____, ____)

P: (____, ____)

SRB
275

② Solve. You may use fraction circle pieces to help you.

a. $\frac{1}{5} + \frac{3}{10} =$ _____

b. $\frac{1}{5} + \frac{1}{10} =$ _____

c. $\frac{2}{3} + \frac{1}{6} =$ _____

SRB
166, 189

③ Write a multiplication problem for which your estimate might be:

$300 \times 70 = 21,000$

_____ × _____ = ?

Solve your problem.

SRB
83,
100-104

④ Round to the nearest hundredth.

a. $67.467 =$ _____

b. $9.017 =$ _____

c. $43.284 =$ _____

d. $16.107 =$ _____.

e. $5.658 =$ _____

SRB
124-127

⑤ **Writing/Reasoning** Explain your strategy for rounding a decimal to the nearest hundredth in Problem 4.

_____

_____

_____

_____

_____

SRB
124-127

149

Finding Areas of New Floors

Several rooms at Westview School will have new tile floors installed next year. Each tile is 1 square yard. For each room, find:

a. the number of tiles needed to cover the floor

b. the area of the floor in square yards

1 Music Room

$7\frac{1}{2}$ yd

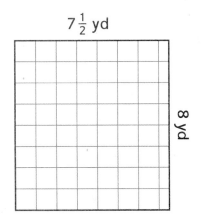

8 yd

a. Number of tiles: _____

b. Area: _____ square yards

2 Office

6 yd

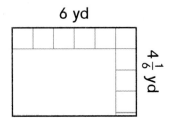

$4\frac{1}{6}$ yd

a. Number of tiles: _____

b. Area: _____ square yards

3 Cafeteria

12 yd

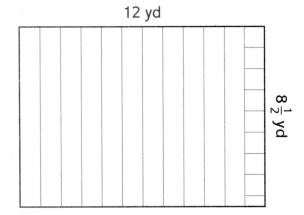

$8\frac{1}{2}$ yd

a. Number of tiles: _____

b. Area: _____ square yards

4 Art Room

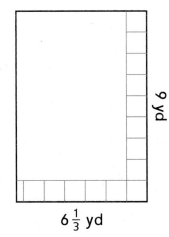

9 yd

$6\frac{1}{3}$ yd

a. Number of tiles: _____

b. Area: _____ square yards

Math Boxes

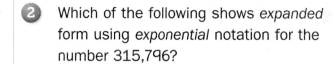

1 Write <, >, or =.

   a. 0.90 _____ 0.89

   b. 3.52 _____ 3.8

   c. 6.91 _____ 6.3

   d. 4.05 _____ 4.2

   e. 0.38 _____ 0.5

SRB
121-123

2 Which of the following shows *expanded* form using *exponential* notation for the number 315,796?

Fill in the circle next to the best answer.

  (A) 300,000 + 10,000 + 5,000 + 700 + 90 + 6

  (B) 3 × 100,000 + 1 × 10,000 + 5 × 1,000 + 7 × 100 + 9 × 10 + 6 × 1

  (C) $3 \times 10^5 + 1 \times 10^4 + 5 \times 10^3 + 7 \times 10^2 + 9 \times 10^1 + 6 \times 10^0$

SRB
68-70

3 Solve.

   a. $\frac{1}{3}$ of 30 = _____

   b. $\frac{1}{8}$ of 16 = _____

   c. $\frac{1}{5}$ of 25 = _____

SRB
195

4 How would you write 54.279 in words? Choose the best answer.

  ( ) fifty-four and two hundred seventy-nine thousandths

  ( ) fifty-four and two hundred seventy-ninths

  ( ) fifty-four and two hundred seventy-nine hundredths

SRB
117

5 **Writing/Reasoning** Write a number story that could be modeled by Problem 3a.

_____

_____

_____

_____

_____

SRB
195

Math Boxes

Math Boxes

1 How many unit squares cover the rectangle?

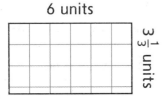

6 units

$3\frac{1}{3}$ units

_____ unit squares

What is the area of the rectangle?

Area = _____

SRB
224-225

2 Write 4 multiples of 11.

_____

_____

SRB
72

3 $\frac{3}{4} - \frac{1}{10}$ _____ $\frac{3}{8} + \frac{1}{12}$

Choose the best answer.

 >

 <

 =

SRB
181-182

4 Write all the factors of 26.

_____

_____

SRB
73

5 Circle True or False.

a. $\frac{1}{3} = \frac{3}{12}$ True   False

b. $\frac{7}{7} = \frac{11}{11}$ True   False

c. $\frac{2}{5} = \frac{6}{15}$ True   False

SRB
166,
168-170

6 Solve.

a. What is $\frac{1}{3}$ of 12? _____

b. What is $\frac{1}{5}$ of 10? _____

SRB
195

Rectangular Prism Patterns

Rectangular Prism A pattern

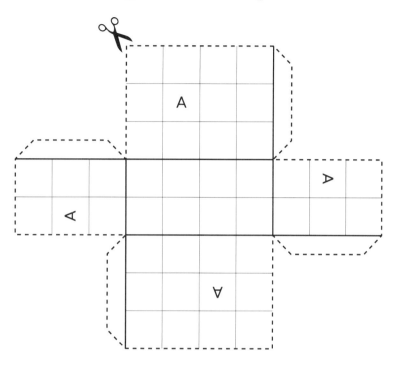

Rectangular Prism B pattern

Rectangular Prism C pattern

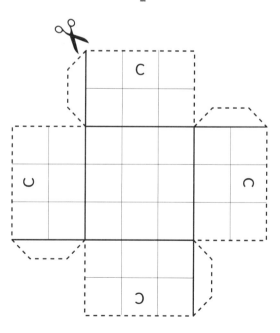

More Rectangular Prism Patterns

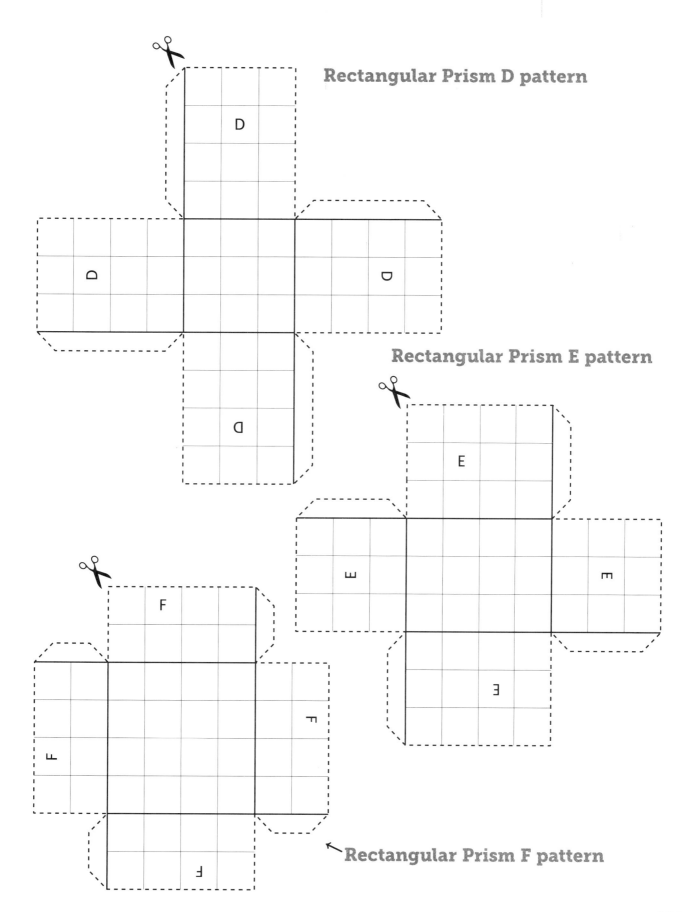

Rectangular Prism D pattern

Rectangular Prism E pattern

Rectangular Prism F pattern

AS2

Prism Pile-Up Cards

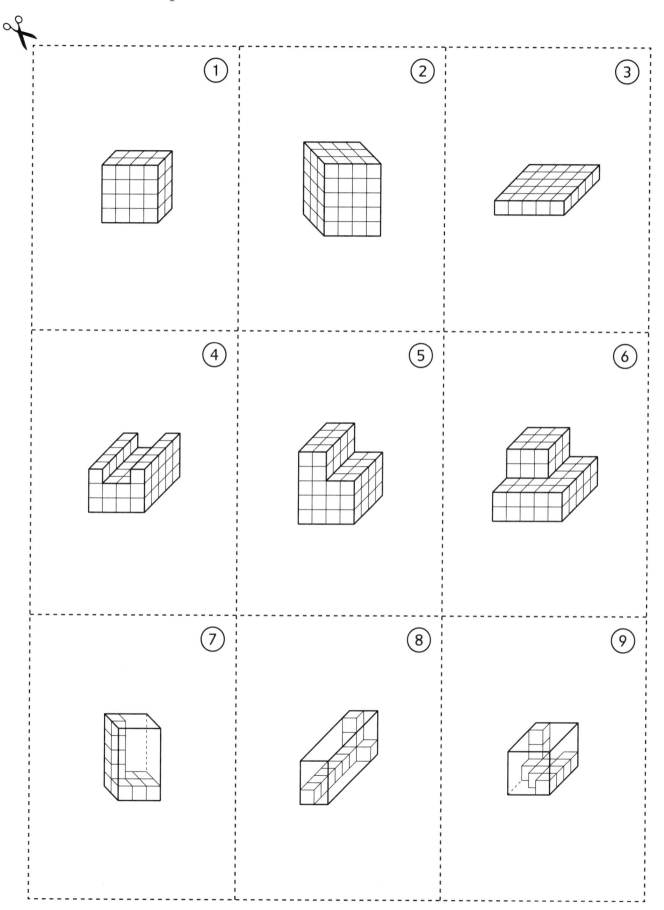

AS3

Prism Pile-Up Cards (continued)

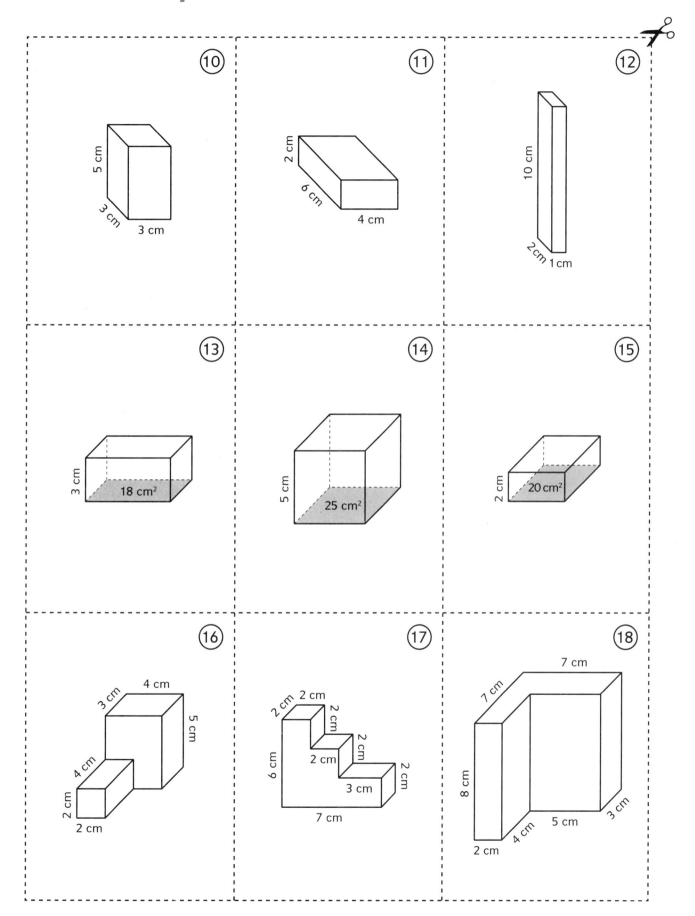

Fraction Circle Pieces 1

Red

Pink

Orange

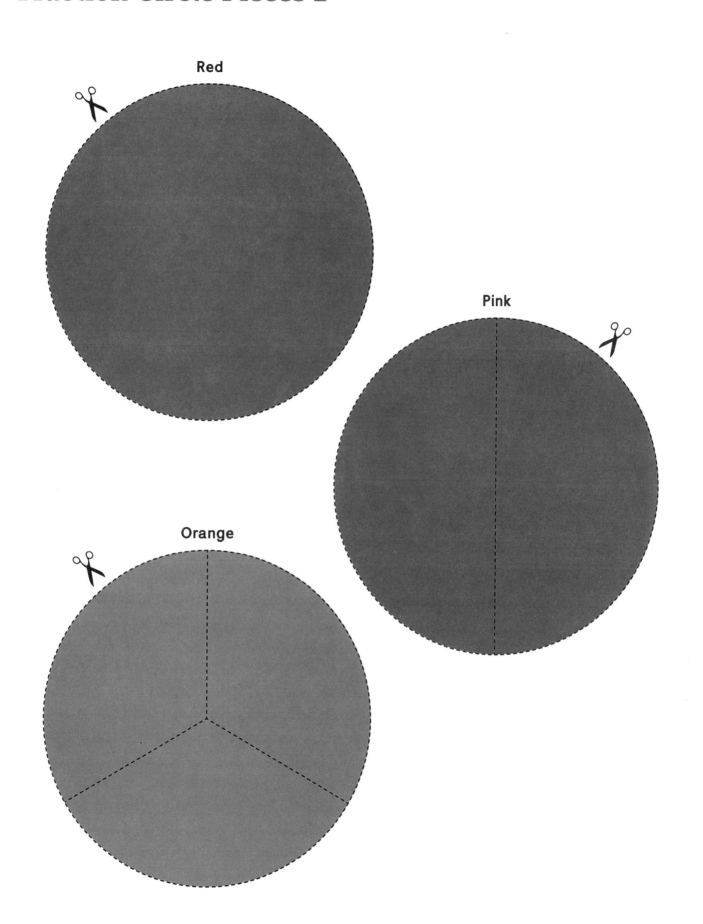

AS5

Fraction Circle Pieces 2

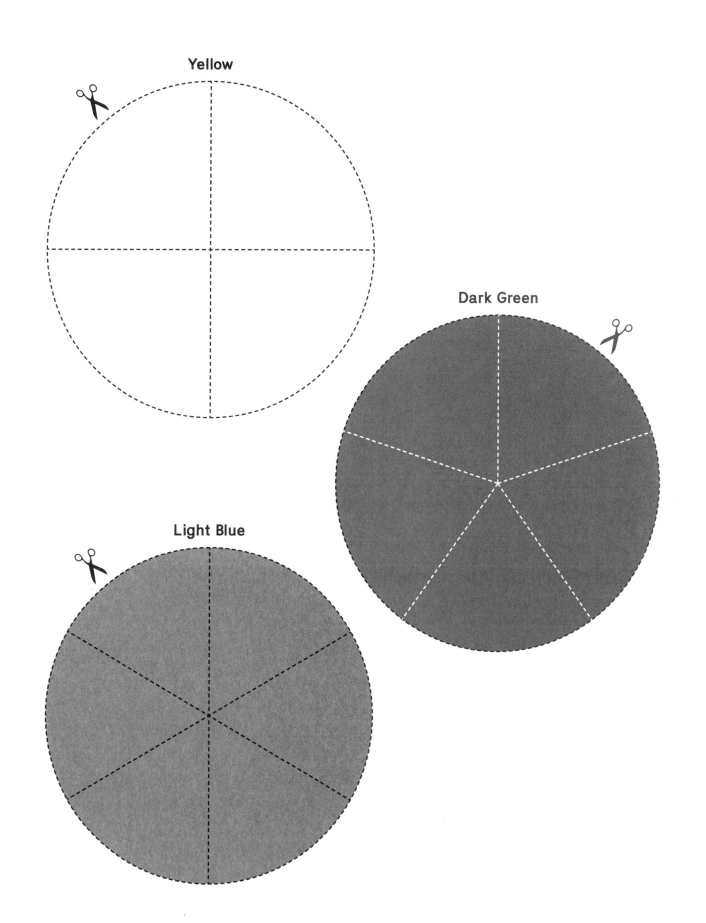

Yellow

Dark Green

Light Blue

Fraction Circle Pieces 3

Dark Blue

Purple

Light Green

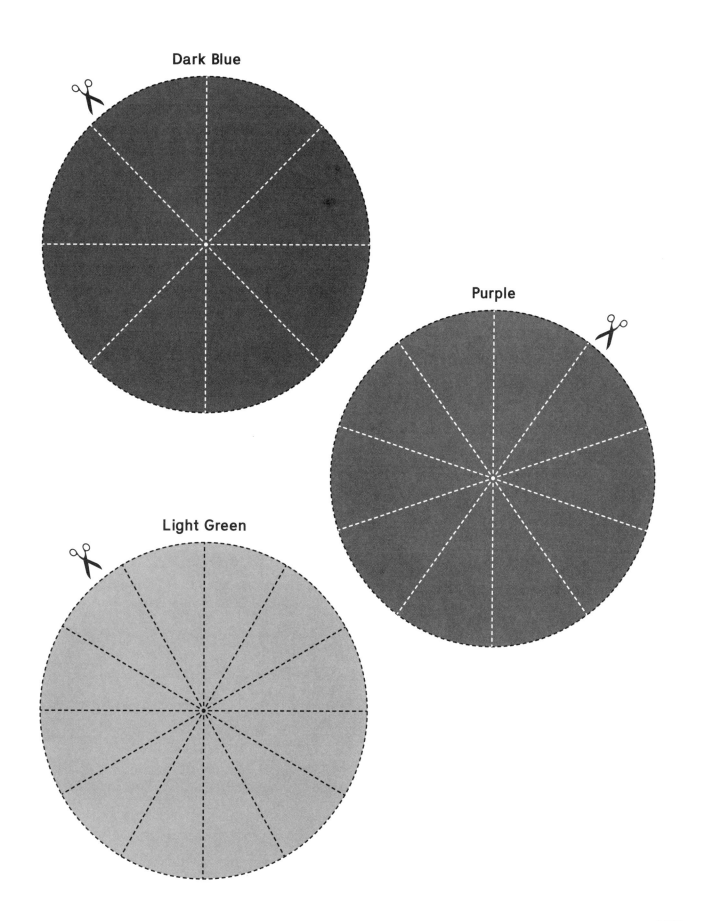

Fraction Cards 1

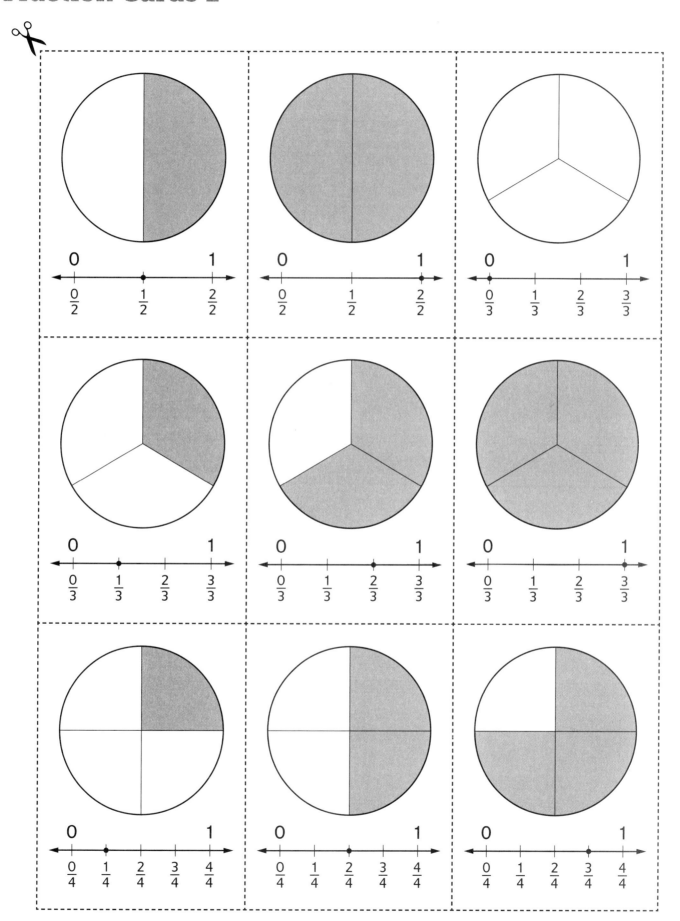

AS8

Fraction Cards 1

$$\frac{0}{3}$$ $$\frac{2}{2}$$ $$\frac{1}{2}$$

$$\frac{3}{3}$$ $$\frac{2}{3}$$ $$\frac{1}{3}$$

$$\frac{3}{4}$$ $$\frac{2}{4}$$ $$\frac{1}{4}$$

Fraction Cards 2

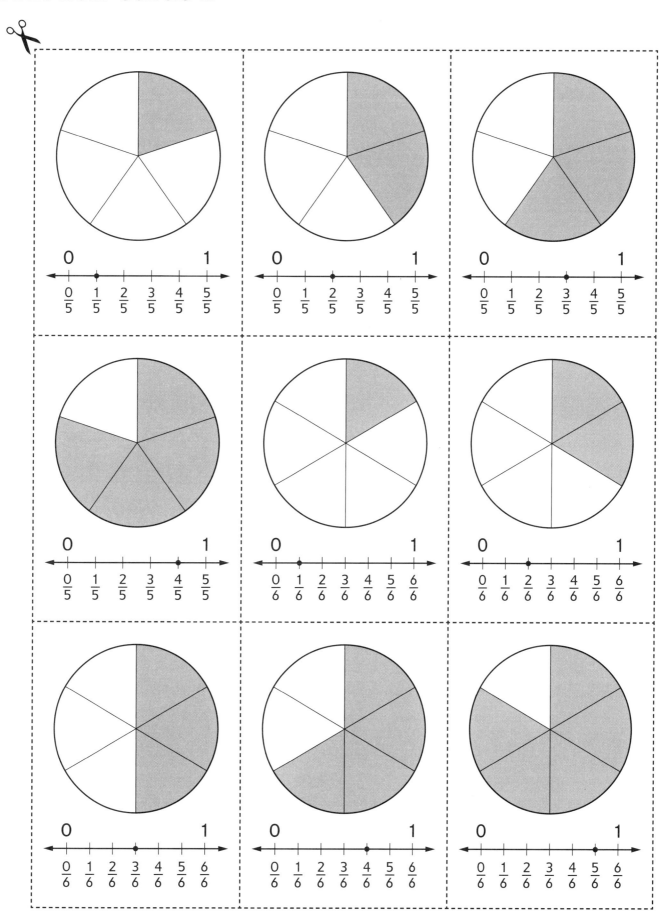

Fraction Cards 2

$$\frac{3}{5}$$

$$\frac{2}{5}$$

$$\frac{1}{5}$$

$$\frac{2}{6}$$

$$\frac{1}{6}$$

$$\frac{4}{5}$$

$$\frac{5}{6}$$

$$\frac{4}{6}$$

$$\frac{3}{6}$$

Fraction Cards 3

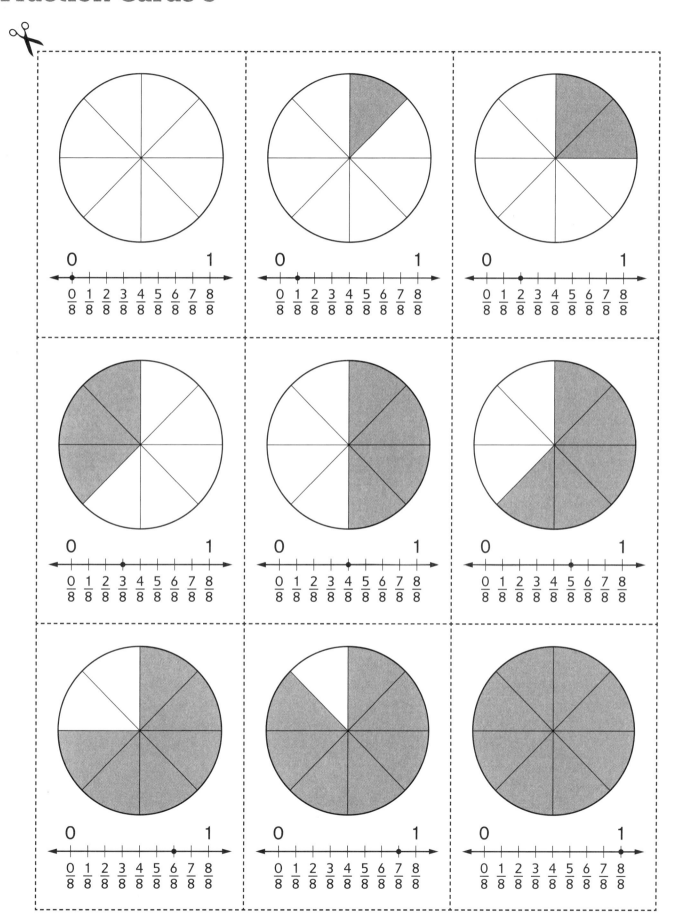

Fraction Cards 3

$$\frac{2}{8} \qquad \frac{1}{8} \qquad \frac{0}{8}$$

$$\frac{5}{8} \qquad \frac{4}{8} \qquad \frac{3}{8}$$

$$\frac{8}{8} \qquad \frac{7}{8} \qquad \frac{6}{8}$$

Fraction Cards 4

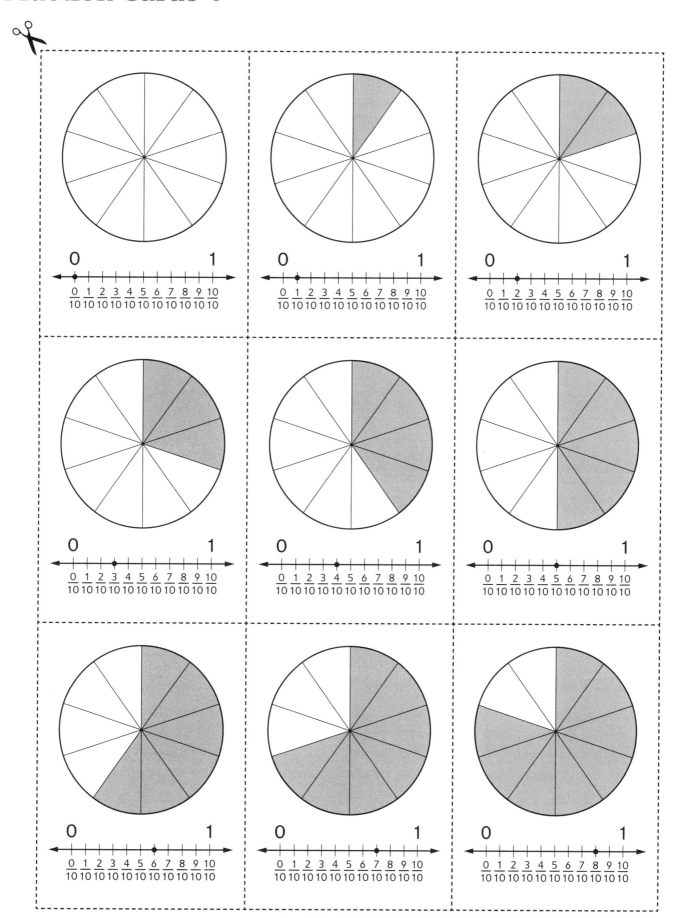

AS11

Fraction Cards 4

$$\frac{2}{10} \qquad \frac{1}{10} \qquad \frac{0}{10}$$

$$\frac{5}{10} \qquad \frac{4}{10} \qquad \frac{3}{10}$$

$$\frac{8}{10} \qquad \frac{7}{10} \qquad \frac{6}{10}$$

Fraction Cards 5

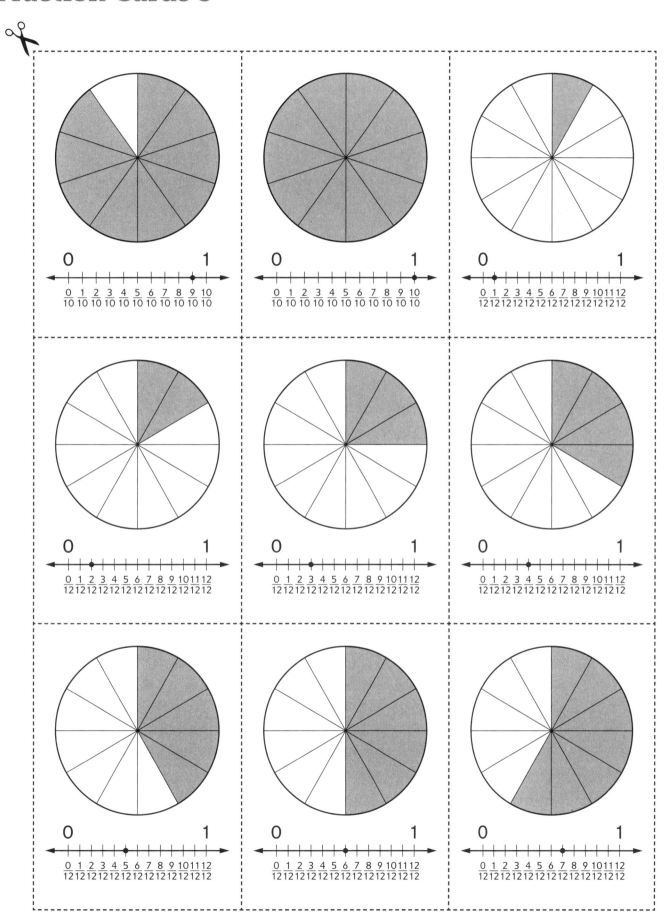

Fraction Cards 5

$$\frac{1}{12}$$ $$\frac{10}{10}$$ $$\frac{9}{10}$$

$$\frac{4}{12}$$ $$\frac{3}{12}$$ $$\frac{2}{12}$$

$$\frac{7}{12}$$ $$\frac{6}{12}$$ $$\frac{5}{12}$$

Fraction Cards 6

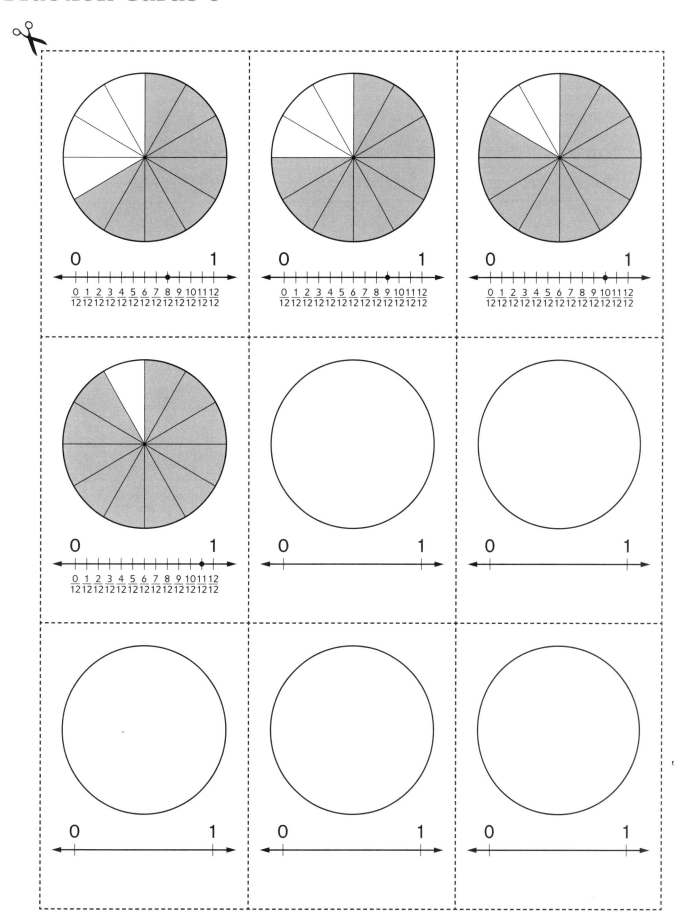

$$\frac{10}{12} \qquad \frac{9}{12} \qquad \frac{8}{12}$$

$$\frac{11}{12}$$

Fraction Of Fraction Cards (Set 1)

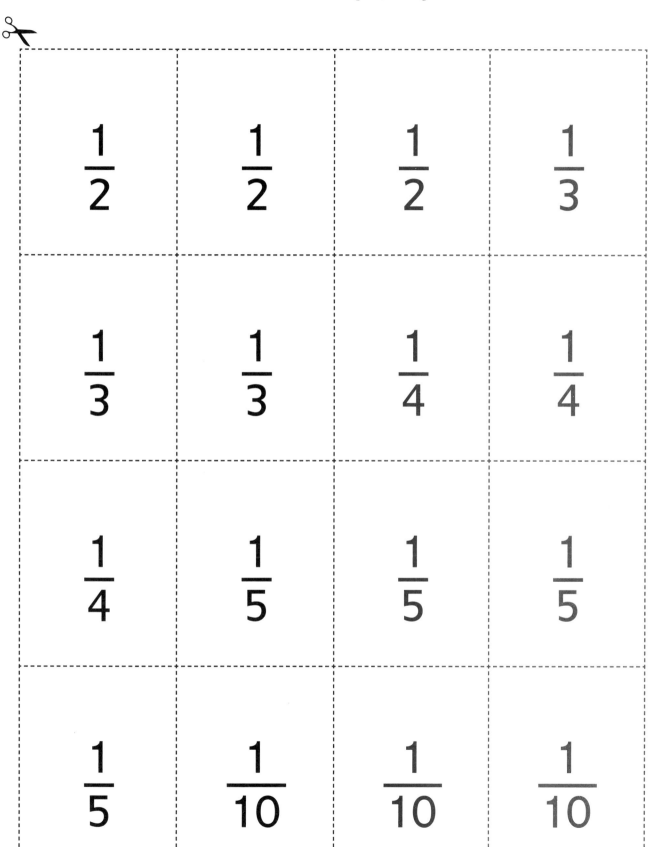

Fraction Of Whole Cards

✂

3 20 15	4 21 30	5 12 20	6 28 40
8 27 20	10 32 24	12 30 25	15 36 20
18 36 10	20 4 3	21 30 24	25 6 40
28 35 30	30 32 15	36 20 24	40 18 25

Blank Cards